Lecture Notes in Statistics　　96
Edited by S. Fienberg, J. Gani, K. Krickeberg,
I. Olkin, and N. Wermuth

UNIVERSITY OF MAINE

Raymond H. Fogler Library

Ibrahim Rahimov

Random Sums and Branching Stochastic Processes

Springer-Verlag
New York Berlin Heidelberg London Paris
Tokyo Hong Kong Barcelona Budapest

Ibrahim Rahimov
The Institute of Mathematics
The Academy of Sciences of the Republic of Uzbekistan
Hodjaev Street, 29,
Tashkent, 700142, Uzbekistan

Department of Statistics
Middle East Technical University
06531, Ankara, Turkey

Library of Congress Cataloging-in-Publication Data Available
Printed on acid-free paper.

© 1995 Springer-Verlag New York, Inc.
All rights reserved. This work may not be translated or copied in whole or in part without the written permission of the publisher (Springer-Verlag New York, Inc., 175 Fifth Avenue, New York, NY 10010, USA), except for brief excerpts in connection with reviews or scholarly analysis. Use in connection with any form of information storage and retrieval, electronic adaptation, computer software, or by similar or dissimilar methodology now known or hereafter developed is forbidden.
The use of general descriptive names, trade names, trademarks, etc., in this publication, even if the former are not especially identified, is not to be taken as a sign that such names, as understood by the Trade Marks and Merchandise Marks Act, may accordingly be used freely by anyone.

Camera ready copy provided by the editor.
Printed and bound by Braun-Brumfield, Ann Arbor, MI.
Printed in the United States of America.

9 8 7 6 5 4 3 2 1

ISBN 0-387-94446-X   Springer-Verlag New York Berlin Heidelberg

TO THE MEMORY OF MY

FATHER AND MY MOTHER

Uzbekistan is one of the Central Asiatic Republics of the former Soviet Union. The ancient Uzbek cities Samarkand, Bukhara, Khiva and Tashkent have long been major attractions. They have over 4000 architectural monuments, many of them under UNESCO protection. Uzbekistan is the mother country of M. Khorezmi (Al-Goritm), Abu Raikhon Beruni and Mirzo Ulugbeg, famous mathematicians and astronomers of the Middle Ages. Ibn Sina (Avitsenna) the great medical scientist, philosopher and poet was born here.

CONTENTS

INTRODUCTION	1
CHAPTER I. SUMS OF A RANDOM NUMBER OF RANDOM VARIABLES.	5
§1.1. **Sampling sums of dependent variables and mixtures of infinitely divisible distributions.**	5
§1a. Sums of a random number of random variables.	7
§1b. Multiple sums of dependent random variables.	9
§1c. Sampling sums from a finite population.	14
§1.2. **Limit theorems for a sum of randomly indexed sequences.**	18
§2a. Sufficient conditions.	18
§2b. Necessary and sufficient conditions.	21
§2c. An application.	25
§1.3. **Necessary and sufficient conditions and limit theorems for sampling sums.**	27
§3a. Convergence theorems.	27
§3b. The rate of convergence.	35
CHAPTER II. BRANCHING PROCESSES WITH GENERALIZED IMMIGRATION.	44
§2.1. **Classical models of branching processes.**	44
§1a. Bellman-Harris processes.	45
§1b. Moments and extinction probabilities.	46
§1c. Asymptotics of non-extinction probability and exponential limit distribution.	48
§1d. Branching processes with stationary immigration.	51
§1e. Continuous time branching processes with immigration.	54
§2.2. **General branching processes with reproduction dependent immigration.**	58
§2a. The model.	59
§2b. The main theorem.	62
§2c. The proof of the main theorem.	64
§2d. Applications of the main theorem.	71
§2.3. **Discrete time processes.**	76

§3a. The model.	76
§3b. Limit theorems for discrete time processes.	78
§3c. Some examples.	84
§3d. Randomly stopped immigration.	87
§2.4. **Convergence to Jirina processes and transfer theorems for branching processes.**	92
§4a. The model.	92
§4b. The main theorem and corollaries.	94
§4c. The proof of the main theorem.	97
CHAPTER III. BRANCHING PROCESSES WITH TIME-DEPENDENT IMMIGRATION.	105
§3.1. **Decreasing immigration.**	105
§1a. The main theorem.	106
§1b. The proof of the main theorem.	118
§1c. State-dependent immigration.	122
§3.2. **Increasing immigration.**	124
§2a. The process with infinite variance.	124
§2b. The process with finite variance.	132
§3.3. **Local limit theorems.**	136
§3a. Occupation of an increasing state.	136
§3b. Occupation of a fixed state.	154
CHAPTER IV. THE ASYMPTOTIC BEHAVIOR OF FAMILIES OF PARTICLES IN BRANCHING PROCESSES.	156
§4.1. **Sums of dependent indicators.**	157
§1a. Sums of functions of independent random variables.	157
§1b. Sampling sums of dependent indicators.	163
§4.2. **Family of particles in critical processes.**	167
§2a. The model.	167
§2b. Limit theorems.	168
§4.3. **Families of particles in supercritical and subcritical processes.**	177
§3a. Supercritical processes.	177
§3b. Subcritical processes.	184
REFERENCES.	184
INDEX.	194

§1.2. Limit Theorems for a Sum of Randomly Indexed Sequences

$$P\left\{\bigcup_{k=1}^{[(1+\delta)m]} \{|\nu(k,t)-l(k,t)|>\delta_o l(k,t)\}\right\}$$

$$+ P\left\{\sum_{k=1}^{[(1+\delta)m]} \max_{i\in\Delta_{\delta_o}(l(k,t))} |Z_i(k,t) - Z_{l(k,t)}(k,t)|>\varepsilon\right\}.$$

According to the condition (I) there exist $\delta=\delta_1$ and t_1 such that the first term is smaller than ε for any $\delta_o>0$ and $t>t_1$. It follows from the condition (II) that there exist $\delta=\delta_2$, δ_o and t_2 such that the second term is less than ε for $t>t_2$. Then for $\delta=\delta_3=\min(\delta_1,\delta_2)$ and $t>\max(t_1,t_2)$ both of inequalities are true, that is,

$$P\left\{|W_3(t)|>\varepsilon;\ |\eta(t)-m(t)|<\delta_3 m(t)\right\} \leq 2\varepsilon.$$

Hence $W_3(t) \xrightarrow{P} 0$ as $t\to\infty$. Thus, we obtain the second part of the lemma from relation (2.3).

§2b. Necessary and Sufficient Conditions.

Let $\varphi(x)$, $x\in R$, be a non-decreasing positive function; $l(x,t)$ are such that $\min_{1\leq k\leq m} l(k,t)\to\infty$ as $t\to\infty$ and let $L(t)$, $\theta(t)$, $I(t)$, $J(t)$ be $m(t)$-dimensional vectors with coordinates $l(k,t)$, $\nu(k,t)$, i_k, j_k, $k=1,2,\ldots,m(t)$, respectively. Denote the vector $(\varphi(x_1),\ldots,\varphi(x_m))$ by $\varphi(X)$ for $X=(x_1,\ldots,x_2)$. For vectors X and $Y=(y_1,\ldots,y_m)$ inequality $X<Y$ means $x_i<y_i$, $i=1,\ldots,m$, and $X\leq Y$ means $x_i\leq y_i$, $i=1,\ldots,m$.

Denote by L the family of indexes $L=\{l(k,t),\ (k,t)\in N^2\}$,

$$D_t(\delta) = \{I(t)\in N^m:\ |\varphi(I(t))-\varphi(L(t))|<\delta\varphi(L(t))\}.$$

If for any $\delta>0$

$$\lim_{t\to\infty} P\{|\varphi(\theta(t))-\varphi(L(t))|<\delta\varphi(L(t))\}=1,$$

we shall say that the family of r.v. $\nu=\{\nu(k,t),(k,t)\in N^2\}$ belongs to the set $K_L(\varphi,m)$.

We introduce now an extended Anskombe condition: for any $\varepsilon>0$ there exists $\delta>0$ such that

22   I. Sums of a Random Number of Random Variables

$$\limsup_{t\to\infty} P\left\{\max_{I(t)\in D_t(\delta)} \left|\sum_{k=1}^{m(t)} (Z_{i_k}(k,t) - Z_{1(k,t)}(k,t))\right| \geq \varepsilon\right\} < \varepsilon. \quad (2.5)$$

The following theorem gives necessary and sufficient conditions for a random change of indexes in $R(t)$ for the index set $K_1(\varphi,m)$.

Theorem 2.1. 1°. The difference $V(t) - R(t) \xrightarrow{P} 0$ as $t\to\infty$ for all $\nu \in K_L(\varphi,m)$ if and only if condition (2.5) is fulfilled.

2°. The following statements are equivalent:

a) the family Z satisfies (2.5) for some $\{k(k,t)\}$ and $R(t) \xrightarrow{d} V$ as $t\to\infty$;

b) variable $V(t) \xrightarrow{d} V$ as $t\to\infty$ for all $\nu \in K_L(\varphi,m)$.

Proof. We prove p.1°. The proof of the sufficiency of the condition (2.5) is just like that of lemma 2.1.

Let us prove the necessity. Let

$$V(t) - R(t) \xrightarrow{P} 0, \quad t\to\infty, \quad (2.6)$$

for any family ν from the set $K_L(\varphi,m)$, and condition (2.5) is not satisfied. Then, for any strongly decreasing sequence of positive numbers δ_n, there exist $\varepsilon > 0$ and a sequence of positive integers $\{t_n, t_n \to \infty\}$ such that

$$P\left\{\max_{I(t_n)\in D_{t_n}(\delta_n)} \left|\sum_{k=1}^{m(t_n)} (Z_{i_k}(k,t_n) - Z_{1(k,t_n)}(k,t_n))\right| \geq \varepsilon\right\} \geq \varepsilon \quad (2.7)$$

for any $n \in N$. Let the strongly decreasing function $\psi(x)$ be such that

$$\psi(\max_{1\leq k\leq m(t_n)} 1(k,t)) = \delta_n \quad \text{for} \quad t_{n-1} < t \leq t_n.$$

Then, from (2.5) we obtain:

$$\liminf_{n\to\infty} P\left\{\max_{I(t_n)\in D_{t_n}} \left|\sum_{k=1}^{m(t_n)} (Z_{i_k}(k,t) - Z_{1(k,t_n)}(k,t_n))\right| \geq \varepsilon\right\} > 0,$$

$$D_t = \left\{I(t) \in N^{m(t)} : \left|\frac{\varphi(i_k)}{\varphi(1(k,t))} - 1\right| < \varphi(1(k,t)), \; k=1,\ldots,m(t)\right\}.$$

The last statement shows that the relation

$$\max_{I(t)\in D_t} \left|\sum_{k=1}^{m(t)} (Z_{i_k}(k,t) - Z_{1(k,t)}(k,t))\right| \xrightarrow{P} 0, \quad (2.8)$$

§1.2. Limit Theorems for a Sum of Randomly Indexed Sequences   23

is not realized for the function $\psi(x)$.

Since the function $\psi(x)$ is decreasing and $\max\limits_{1\leq k\leq m(t)} l(k,t)\to\infty$ as $t\to\infty$, then

$$P\left\{\bigcup_{k=1}^{m(t)}\left\{\left|\frac{\varphi(\nu(k,t))}{\varphi(l(k,t))}-1\right|\geq \psi(l(k,t))\right\}\right\}$$

$$\geq P\left\{\max_{1\leq k\leq m}\left\{\left|\frac{\varphi(\nu(k,t))}{\varphi(l(k,t))}-1\right|>\delta\right\}\right\} \qquad (2.9)$$

for any $\delta>0$ and for any sufficiently large t (it is so, for example, if t is such that $\delta>\psi(\min\limits_{1\leq k\leq m(t)} l(k,t))$). It follows from (2.9) that, if the family ν is such that

$$P\left\{\bigcup_{k=1}^{m(t)}\left\{\left|\frac{\varphi(\nu(k,t))}{\varphi(l(k,t))}-1\right|>\psi(l(k,t))\right\}\right\}\longrightarrow 0, \qquad (2.10)$$

then $\nu\in K_L(\varphi,m)$. Hence (2.6) is true for any family ν which satisfies (2.10) according to the initial assumption.

Now we show that, if (2.6) is true for any family ν satisfying (2.10), then (2.8) is necessarily satisfied for Z. Construct the index family ν in the following way:

We put

$$\theta(t) = \left\{J(t)\in N^{m(t)}: \left|\sum_{k=1}^{m(t)}(Z_{j_k}(k,t)-Z_{l(k,t)}(k,t))\right|\right.$$

$$= \max_{I(t)\in D_t}\left|\sum_{k=1}^{m(t)}(Z_{i_k}(k,t)-Z_{l(k,t)}(k,t))\right|\right\},$$

for any t. It is clear that relations (2.6) and (2.8) are equivalent for this index family. On the other hand, $\theta(t)\in D_t$ with probability 1 for any t. So, it follows from the relation

$$P\left\{\theta(t)\in D_t\right\} = 1 - P\left\{\bigcup_{k=1}^{m(t)}\left\{\left|\frac{\varphi(\nu(k,t))}{\varphi(l(k,t))}-1\right|>\psi(l(k,t))\right\}\right\}$$

that the family ν satisfies (2.10). Hence, for (2.6) to be correct it is necessary that (2.8) is true. But (2.8) must not be fulfilled according to the initial assumption. This contradiction proves part 1 of the theorem.

Let us prove part 2. It follows from the first part of the theorem that a)⇒b). We prove that b) implies a). For convergence $V(t) \xrightarrow{d} V$ as $t \to \infty$ for all families $\nu \in K_L(\varphi, m)$, it is necessary convergence $R(t) \xrightarrow{d} V$ as $t \to \infty$. Let $R(t) \xrightarrow{d} V$ and condition (2.5) is true. Then, for any strongly decreasing sequence of positive numbers $\{\delta_n\}$, there exist $\varepsilon > 0$ and sequence $\{t_n, t_n \to \infty\}$ such that (2.7) holds.

Consider the decomposition $R = \bigcup_{i=-\infty}^{\infty} B_i$ of the number line by non-crossing sets

$$B_i = \left(\frac{\varepsilon i}{4}, \frac{\varepsilon(i+1)}{4}\right).$$

It is clear that there exists some finite subset E of the set of integers such that

$$3P\left\{V \notin \bigcup_{i \in E} B_i\right\} \leq \varepsilon. \qquad (2.11)$$

It follows from the condition $R(t) \xrightarrow{d} V$, $t \to \infty$, that there exists natural number N_0 such that

$$P\left\{R(t_n) \notin \bigcup_{i \in E} B_i\right\} \leq \frac{3}{2} P\left\{V \notin \bigcup_{i \in E} B_i\right\}, \qquad (2.12)$$

for any $n \geq N_0$. From relations (2.11), (2.12) and inequality

$$\varepsilon \leq P\left\{\max_{I(t_n) \in D_{t_n}(\delta_n)} \left| \sum_{k=1}^{m(t_n)} (Z_{i_k}(k, t_n) - Z_{1(k, t_n)}(k, t_n)) \right| > \varepsilon \right\}$$

$$\leq P\left\{\max_{I(t_n) \in D_{t_n}(\delta_n)} \left| \sum_{k=1}^{m(t_n)} (Z_{i_k}(k, t_n) - Z_{1(k, t_n)}(k, t_n)) \right| > \varepsilon, R(t_n) \in \bigcup_{i \in E} B_i \right\}$$

$$+ P\left\{R(t_n) \notin \bigcup_{i \in E} B_i\right\}$$

we obtain that

$$P\left\{T_n > \varepsilon, R(t_n) \in \bigcup_{i \in E} B_i\right\} \geq \varepsilon/2, \qquad (2.13)$$

for $n \geq N_0$, where

$$T_n = \max_{I(t_n) \in D_{t_n}(\delta_n)} \left| \sum_{k=1}^{m(t_n)} (Z_{i_k}(k, t_n) - Z_{1(k, t_n)}(k, t_n)) \right|.$$

Since B_i are non-crossing sets,

$$P\left\{T_n > \varepsilon, R(t_n) \in \bigcup_{i \in E} B_i\right\} \leq |E| \max_{i \in E} P\left\{T_n > \varepsilon, R(t_n) \in B_i\right\},$$

§1.2. Limit Theorems for a Sum of Randomly Indexed Sequences

where $|E|$ is the number of elements of the set E. Hence,

$$\max_{i \in E} P\{T_n > \varepsilon, R(t_n) \in B_i\} \geq \frac{\varepsilon}{2|E|}, \quad n \geq N_o. \quad (2.14)$$

Let the maximum on the left hand side of (2.14) be achieved on a set B. Define $\theta(t_n)$ in the following way. We put $\theta(t_n, \omega) = J(t_n)$ if ω belongs to

$$\left\{ \left| \sum_{k=1}^{m(t_n)} (z_{j_k}(k, t_n) - z_{1(k,t_n)}(k, t_n)) \right| = T_n, \quad R(t_n) \in B \right\}$$

and $\theta(t_n, \omega) = L(t_n)$ elsewhere. Then it is clear that,

$$\{R(t_n) \in B, T_n > \varepsilon\} = \{R(t_n) \in B, |V(t_n) - R(t_n)| > \varepsilon\}$$

$$\subseteq \{R(t_n) \in B, V(t_n) \notin B\}. \quad (2.15)$$

Thus, from (2.14) and (2.15), we obtain that

$$P\{V(t_n) \notin B\} \geq P\{V(t_n) \notin B, R(t_n) \notin B\} + \frac{\varepsilon}{2|E|}, \quad n \geq N_o. \quad (2.16)$$

Since $\{R(t_n) \notin B\} \subset \{V(t_n) \notin B\}$ according to the construction of $\theta(t_n)$, we have from (2.16):

$$P\{V(t_n) \notin B\} \geq P\{R(t_n) \notin B\} + \frac{\varepsilon}{2|E|}, \quad n \geq N_o. \quad (2.17)$$

Inequality (2.17) shows that the variable $V(t)$ cannot converge to V in distribution. This contradiction proves that b) implies a).

§2c. An Application.

We now consider an example of a scheme which satisfies some of the conditions of Lemma 2.1. Let us denote

$$S_t = \sum_{k=1}^{m(t)} \sum_{j=1}^{\nu(k,t)} \xi_{jk}(t), \quad S_t^* = \sum_{k=1}^{m(t)} \sum_{j=1}^{l(k,t)} \xi_{jk}(t),$$

where $\{\xi_{jk}(t)\}$ are independent non-negative random variables, having common distribution for different j and $M\xi_{1k}(t) < \infty$ for any $(k,t) \in \mathbb{N}^2$,

$$\bar{a}(k,t) = M\xi_1(k,t), \quad \bar{S}_t = \sum_{k=1}^{m(t)} \nu(k,t)\bar{a}(k,t).$$

Theorem 2.2. Suppose that $S_t^* / MS_t^* \xrightarrow{d} V$ as $t \to \infty$ for some family $\{l(k,t), k \in N\}$. If condition (I) is satisfied for $M=1$, then

$$\frac{S_t}{\bar{S}_t} \xrightarrow{d} V, \quad \frac{S_t}{MS_t^*} \xrightarrow{d} V. \qquad (2.18)$$

Proof. Let us prove the first statement. We consider

$$\frac{S_t}{\bar{S}_t} = \frac{MS_t^*}{\bar{S}_t} \sum_{k=1}^{m(t)} Z_{\nu(k,t)}(k,t), \qquad (2.19)$$

where

$$Z_{\nu(k,t)}(k,t) = \frac{1}{MS_t^*} \sum_{j=1}^{\nu(k,t)} \xi_{jk}(t).$$

We show that condition (II) is satisfied with $M=1$ for the sum on the right-hand side of (2.19). Consider the partition

$$\Delta_\delta(l(k,t)) = \Delta_\delta^{(1)}(l(k,t)) \cup \Delta_\delta^{(2)}(l(k,t)), \quad \delta \in (0,1),$$

where the first set consists of $i \in \Delta_\delta(l(k,t))$ such that $i \le l(k,t)$ and the second set consists of $i \in \Delta_\delta(l(k,t))$ such that $i > l(k,t)$. Then we obtain that

$$R = P\left\{\sum_{k=1}^{m_t} \max_{i \in \Delta_\delta(l(k,t))} \left|(Z_i(k,t) - Z_{l(k,t)}(k,t)\right| > \varepsilon\right\}$$

$$\le P\left\{\sum_{k=1}^{m_t} \sum_{j=1}^{2} \max_{i \in \Delta_\delta^{(i)}(l(k,t))} \left|(Z_i(k,t) - Z_{l(k,t)}(k,t)\right| > \varepsilon\right\}, \qquad (2.20)$$

for any $\varepsilon > 0$. Since $\xi_{jk}(t)$ are non-negative,

$$\max_{i \in \Delta_\delta^{(1)}(l(k,t))} \left|(Z_i(k,t) - Z_{l(k,t)}(k,t)\right| = (MS_t^*)^{-1} \sum_{j=[l(k,t)(1-\delta)]+2}^{l(k,t)} \xi_{jk}(t). \quad (2.21)$$

We similarly obtain for the second part:

$$\max_{i\in\Delta_\delta^{(2)}(1(k,t))} \left|(Z_i(k,t) - Z_{1(k,t)}(k,t)\right| = (MS_t^*)^{-1} \sum_{j=[1(k,t)(1-\delta)]+2}^{[1(k,t)(1+\delta)]} \xi_{jk}(t). \quad (2.22)$$

Using (2.21) and (2.22) we obtain from (2.20)

$$R \le P\left\{\sum_{k=1}^{m_t} \sum_{j=r_1}^{r_2} \xi_{jk}(t) > \varepsilon MS_t^*\right\},$$

where

$$r_1 = [1(k,t)(1-\delta)]+2, \quad r_2 = [1(k,t)(1+\delta)].$$

With the help of Chebyshev's inequality we have the inequality $R \le 2\delta/\varepsilon$. This shows that condition (II) is satisfied. Thus, according to Lemma 2.1,

$$\sum_{k=1}^{m_t} z_{\nu(k,t)}(k,t) \xrightarrow{D} V, \quad t\to\infty \quad (2.23)$$

On the other hand, since

$$P\left\{\left|\frac{\bar{S}_t}{MS_t^*} - 1\right| > \delta\right\} \le P\left\{\bigcup_{k=1}^{m(t)} \{|\nu(k,t)-1(k,t)|>\delta 1(k,t)\}\right\},$$

we obtain that under condition (I)

$$\frac{\hat{S}_t}{MS_t^*} \xrightarrow{P} 1. \quad (2.24)$$

The first part of the theorem follows from relations (2.19), (2.23) and (2.24). The proof of the second part we obtain directly from (2.23).

§1.3. NECESSARY AND SUFFICIENT CONDITIONS AND LIMIT THEOREMS FOR SAMPLING SUMS

§3a. Convergence Theorems

In this section we will study the sums considered in §1.1 under some additional assumptions. Let $\{X(k,n), k, n\in N\}$ be a family of **independent** random variables and let $\{\nu(k,n), k, n\in N\}$ be a family of random variables, taking values 0 and 1. As before, we put

$$V_n = \sum_{k=1}^{n} \nu(k,n)X(k,n). \qquad (3.1)$$

In contrast to traditional sampling sums, we allow arbitrary dependence of variables $\{\nu(k,n), k=1,2,\ldots,n\}$.

Below we obtain an approximation for the characteristic function of V_n. Estimate the exactness of this approximation in the case when the populations of random variables are independent. It is easy to verify the fulfillment of the conditions of Theorem 3.1 for the multiple sums considered in §1.1.

We will consider some applications of these results to the theory of branching processes in §3.2. We establish here only two corollaries on the limit distribution of sampling sums with a random sample size in a scheme of sampling without replacement.

We introduce the following notations:

$$a(k,n)=MX(k,n), \quad b(k,n)=DX(k,n), \quad \tilde{S}_n^{(1)} = \sum_{k=1}^{n} \nu(k,n)a(k,n),$$

$$A_n=MV_n, \quad \tilde{S}_n^{(2)} = \sum_{k=1}^{n} \nu(k,n)b(k,n), \quad \Delta_n^2 = M\tilde{S}_n^{(2)}, \quad \sigma_n^2 = D\tilde{S}_n^{(1)},$$

$$B_n^2 = \Delta_n^2 + \sigma_n^2, \quad \Phi_{kn}(t) = Me^{itX(k,n)}, \quad \hat{\Phi}_{kn}(t) = \Phi_{kn}(t)e^{-ita(k,n)}.$$

Denote by $U_n(x)$ and $V_n(x)$ the distributions of variables $(V_n-A_n)/B_n$ and $(\tilde{S}_n-A_n)/\sigma_n$, respectively. Let us denote

$$L_n(\varepsilon) = \frac{1}{B_n^2} \sum_{k=1}^{n} \nu(k,n)\delta_{kn}(\varepsilon B_n),$$

$$\delta_{kn}(x)=M[(X(k,n)-a(k,n))^2, \quad |X(k,n)-a(k,n)|>x],$$

$$\varphi_n(t) = e^{\frac{itA_n}{B_n}} \prod_{k=1}^{n} \Phi_{kn}^{\nu(k,n)}\left(\frac{t}{B_n}\right), \quad \psi_n(t) = e^{it\frac{\tilde{S}_n^{(1)}-A_n}{B_n} - \frac{t^2\tilde{S}_n^{(2)}}{2B_n^2}}.$$

We need the condition

$$\sum_{k=1}^{n} \nu(k,n)\left|1-\hat{\Phi}_{kn}\left(\frac{t}{B_n}\right)\right|^2 \xrightarrow{P} 0, \quad n\to\infty. \qquad (3.2)$$

Theorem 3.1. Suppose that condition (3.2) is satisfied. Then

$$\varphi_n(t) - \psi_n(t) \xrightarrow{P} 0 \qquad (3.3)$$

§1.3. Necessary and Sufficient Conditions

as $n\to\infty$ if and only if the variable $L_n(\varepsilon)$ converges to zero as $n\to\infty$ in probability.

Now we shall mention several corollaries of Theorem 3.1. Let F_0, F_k^n, k, $n\in N$, be σ-algebras determined in §1.1. (see Statement 1.1.2).

Corollary 3.1. If conditions of Theorem 3.1 are satisfied, $\nu(k,n)$, $k,n\in N$, are measurable with respect to $F_{k-1}^{(n)}$ and $\psi_n(t)\xrightarrow{P}\psi(t)$, $n\to\infty$, where $\psi(t)$ is a $\mathcal{F}_0$-measurable random variable for any $t\in R$, $|\psi(t)|>0$ almost everywhere, then

$$Me^{it\frac{V_n - A_n}{B_n}} = M\psi(t) + o(1), \quad n\to\infty.$$

In particular, if $\delta_n^2 = o(\Delta_n^2)$, $\tilde{S}_n^{(2)}/\Delta_n^2 \xrightarrow{P} 1$, $n\to\infty$, then $(V_n - A_n)/\Delta_n$ is asymptotically normal with $(0,1)$ as $n\to\infty$.

Corollary 3.2. Suppose that the conditions of Theorem 3.1 are satisfied, and the vectors $(\nu(1,n),..\nu(n,n))$ and $(X(1,n),...,X(n,n))$ are independent for any n. Then:

1° If $\sigma_n^2 = o(B_n^2)$, $\tilde{S}_n^{(2)}/B_n^2 \xrightarrow{d} \xi$, $n\to\infty$, $F(x) = P\{\xi < x\}$, then

$$U_n(x) = \int_0^\infty \Phi\left(\frac{x}{\sqrt{y}}\right) dF(y) + o(1), \quad n\to\infty;$$

2° if $\sigma_n^2 \sim B_n^2$, $V_n(x) \to V(x)$, $n\to\infty$, then

$$U_n(x) = V(x) + o(1);$$

3° if $\sigma_n^2 \sim \theta B_n^2$, $\theta\in(0,1)$, $V_n(x) \to V(x)$, $\tilde{S}_n^{(2)}/B_n^2 \xrightarrow{P} c\in(0,\infty)$, then

$$U_n(x) = \int_{-\infty}^\infty \Phi\left(\frac{x-u}{\sqrt{c}}\right) dV\left(\frac{u}{\sqrt{\theta}}\right) + o(1), \quad n\to\infty.$$

Let $\{X(k,n), k,n\in N\}$ be a family of certain (dependent in general) random variables, such that the variable $X(k,n)$ is measurable with respect to F_k^n for any pair (k,n), where F_k^n, F_0 are σ-algebras, defined in §1.1, and let

$$f_{kn}(t) = M\left[e^{iX(k,n)t} \middle| F_{k-1}^n\right].$$

Lemma 3.1. Let $\nu(k,n)$ be measurable with respect to F_{k-1}^n for any $k=1,2,\ldots,n$, and

$$\prod_{k=1}^n f_{kn}^{\nu(k,n)}(t) \xrightarrow{P} \varphi(t), \quad n\to\infty,$$

for some $t\in R$, where $\varphi(t)$ is a complex-valued F_0-measurable random variable, whose module is positive almost everywhere. Then

$$M\left[e^{itV_n} \mid F_0\right] \xrightarrow{P} \varphi(t), \quad n\to\infty.$$

Remark 3.1. Such results were obtained by Kubaski (1983) and Klopotovcki (1980) for simple sums by more tedious arguments. We shall deduce Lemma 3.1 from the well-known theorem on the convergence of conditional characteristic functions of semimartingales (see Liptser and Shiryaev, 1986).

Proof of Lemma 3.1. Let us denote

$$V_\tau^{(n)} = \sum_{k=0}^{[n\tau]} \nu(k,n) X(k,n), \quad \tau \in R_+, \quad X(0,n)=0, \quad \nu(0,n)=0.$$

The process $V_\tau^{(n)}$ is representable as a difference of two processes with non-decreasing trajectories:

$$V_\tau^{(n)} = {}_1V_\tau^{(n)} - {}_2V_\tau^{(n)},$$

where

$${}_1V_\tau^{(n)} = \sum_{k=0}^{[n\tau]} \nu_k(n) X^+(k,n), \quad {}_2V_\tau^{(n)} = \sum_{k=0}^{[n\tau]} \nu_k(n) X^-(k,n),$$

$$X^+(k,n) = X(k,n)\wedge 0, \quad X^-(k,n) = -(X(k,n)\wedge 0).$$

Since

$$\mathrm{Var}(V_\tau^{(n)}) = \int_0^\tau |dV_s^{(n)}| = {}_1V_\tau^{(n)} + {}_2V_\tau^{(n)}$$

and the variables $X(k,n)$ are finite, the trajectories of the process $V_s^{(n)}$ have bounded variations in any finite interval $[0,\tau]$. Hence, $(V_\tau^{(n)}, F_{[n\tau]}^{(n)})$ is a semimartingale (Liptser, Shiryaev, 1986; Jakod, Shiryaev, 1987). We now calculate the triplet $T^n = (B^n, C^n, \pi^n)$. We find that the measure of jumps of $V_\tau^{(n)}$ is equal to

$$\mu^{(n)}((0,\tau] \, \Gamma) = \sum_{k=0}^{[n\tau]} \nu(k,n) \chi(X(k,n) \in \Gamma),$$

where Γ is a Borel set of $R_0 = R \setminus \{0\}$. For any Borel set not containing the point zero

$$\{\nu(k,n) X(k,n) \in \Gamma\} = \{\nu(k,n) = 1\} \cap \{X(k,n) \in \Gamma\}.$$

Therefore

1.3. Necessary and Sufficient Conditions

$$\mu^{(n)}((0,\tau]\times\Gamma) = \sum_{k=0}^{[n\tau]} \nu(k,n)\chi\{X(k,n)\in\Gamma\},$$

and the compensator of the measure of jumps is equal to

$$\pi^{(n)}((0,\tau]\times\Gamma) = \sum_{k=0}^{[n\tau]} \nu(k,n)P\left\{X(k,n)\in\Gamma \mid F_{k-1}^n\right\}.$$

Here $F_{-1}^n = \{\emptyset, \Omega\}$. Hence

$$B_\tau^{(n)} = \sum_{k=0}^{[n\tau]} \nu(k,n) M\left[X(k,n)\chi(|X(k,n)|\leq 1) \mid F_{k-1}^n\right].$$

In our case $C_\tau^{(n)} \equiv 0$. Now we calculate the stochastic exponent of the process $V_\tau^{(n)}$. Let

$$\pi_{kn}(x) = P\left\{X(k,n)\leq x \mid F_{k-1}^n\right\}.$$

We obtain (see Liptser and Shiryaev, 1986, p. 256):

$$G_\tau^n(t) = itB_\tau^n + \sum_{k=0}^{[n\tau]} \nu(k,n) \int_{R_0} (e^{itx}-1-itx\chi(|x|\leq 1))\pi_{kn}(dx)$$

$$= \sum_{k=0}^{[n\tau]} \nu(k,n) M\left[(e^{itX(k,n)}-1) \mid F_{k-1}^n\right].$$

Hence, putting $\Delta G_s^n(t) = G_s^n(t) - G_{s-}^n(t)$, by the definition of the stochastic exponent we obtain that

$$\varepsilon_\tau(G_\tau^n)) = e^{G_\tau^n(t)} \prod_{0\leq s\leq \tau} (1+\Delta G_s^m(t))e^{-\Delta G_s^n(t)}$$

$$= \prod_{k=0}^{[n\tau]} [1+\nu(k,n) M\left[(e^{itX(k,n)}-1) \mid F_{k-1}^n\right].$$

It is clear that the random variable under the product sign is equal to $f_{kn}^{\nu(k,n)}(t)$ for any $\omega\in\Omega$. Therefore

$$\varepsilon_\tau(G_\tau^m(t)) = \prod_{k=0}^{[n\tau]} f_{kn}^{\nu(k,n)}(t).$$

Hence by virtue of the conditions of Lemma 3.1,

$$\varepsilon_1(G_1^n(t)) \xrightarrow{P} \varphi(t).$$

Since the function $\varphi(t)$ is F_0-measurable, by virtue of Theorem 4.4.1 of Liptser and Shiryaev (1986) the semimartingale V_τ, having such a stochastic exponent, is a process with F_0-independent increments. Therefore, (Liptser, Shiryaev, 1986, p. 256),

$$\varphi(t) = \varepsilon_1(G_1(t)) = M\left[e^{itV_1} \mid F_0\right].$$

Then, by Theorem 5.1.1 of the same book,

$$M\left[e^{itV_1(n)} \mid F_{n-1}^n\right] \to \varphi(t), \quad n \to \infty.$$

From this relation and from the theorem on convergence under the conditional expectation sign (Shirayev, 1980, p. 232) we get

$$M\left[e^{itV_n} \mid F_0\right] \xrightarrow{P} \varphi(t),$$

as $n \to \infty$. Lemma 3.1 is now proved.

Corollary 3.1 follows from Theorem 3.1 and Lemma 3.1. Corollary 3.2 can be deduced from Corollary 3.1. Now we shall prove Theorem 3.1.

Proof of Theorem 3.1. We represent the function $\varphi_n(t)$ in the form:

$$\varphi_n(t) = \exp\left\{it\,\frac{\tilde{S}_n^{(1)} - A_n}{B_n}\right\} \prod_{k=1}^{n} \left(\hat{\Phi}_{kn}\left(\frac{t}{B_n}\right)\right)^{\nu(k,n)}.$$

It is clear that

$$\operatorname{Re}(\hat{\Phi}_{kn}(t) - 1) = M\cos[t(X(k,n) - a(k,n))] - 1 \leq 0. \tag{3.4}$$

Since $|e^z| = |e^{\operatorname{Re} z}|$,

$$\left|\exp\{\nu(k,n)(\hat{\Phi}_{kn}(t) - 1)\}\right| \leq 1.$$

Therefore, if we use the inequality

$$\left|\prod_k a_k - \prod_k b_k\right| \leq \sum_k |a_k - b_k|, \tag{3.5}$$

which is true for $|a_k| \leq 1$, $|b_k| \leq 1$, we obtain

$$\left|\varphi_n(t) - \exp\left\{\sum_{k=1}^{n} \nu(k,n)\left(\hat{\Phi}_{kn}\left(\frac{t}{B_n}\right) - 1\right) + it\,\frac{\tilde{S}_n^{(1)} - A_n}{B_n}\right\}\right|

1.3. Necessary and Sufficient Conditions

$$\leq \sum_{k=1}^{n} \nu(k,n) \left| \hat{\Phi}_{kn}\left(\frac{t}{B_n}\right) - e^{\hat{\Phi}_{kn}\left(\frac{t}{B_n}\right)-1} \right|.$$

If we take into account (3.4) and use the inequality

$$\left| e^{\alpha} - 1 - \alpha \right| \leq \frac{1}{2} |\alpha|^2, \qquad (3.6)$$

which is true for $\text{Re}\alpha \leq 0$, we obtain that the last sum is less than

$$\sum_{k=1}^{n} \nu(k,n) \left| \hat{\Phi}_{kn}\left(\frac{t}{B_n}\right) - 1 \right|^2.$$

This shows that when $n \to \infty$

$$\varphi_n(t) - \exp\left\{ \sum_{k=1}^{n} \nu(k,n) \left(\hat{\Phi}_{kn}\left(\frac{t}{B_n}\right) - 1 \right) + it\, \frac{\tilde{S}_n^{(1)} - A_n}{B_n} \right\} \xrightarrow{P} 0,$$

under the condition (3.2). Consequently, $\varphi_n(t)$ is representable in the form:

$$\varphi_n(t) = \psi_n(t) \exp\left\{ \sum_{k=1}^{n} \nu(k,n) \rho(k,n) \right\} + \varepsilon_n^{(1)}, \qquad (3.7)$$

where $\varepsilon_n^{(1)} \xrightarrow{P} 0$ as $n \to \infty$ and

$$\rho(k,n) = \hat{\Phi}_{kn}\left(\frac{t}{B_n}\right) - 1 + \frac{t^2 b(k,n)}{2B_n^2}.$$

We shall prove the sufficiency of the condition $L_n(\varepsilon) \xrightarrow{P} 0$ as $n \to \infty$. Putting $\alpha = t(X(k,n) - a(k,n))/B_n$ we represent the function $\rho(k,n)$ as

$$\rho(k,n) = M[e^{i\alpha} - 1 + \alpha^2/2 - i\alpha;\ |\alpha| \leq \varepsilon t]$$
$$+ 2^{-1} M[\alpha^2;\ |\alpha| > \varepsilon t] + M[e^{i\alpha} - 1 - i\alpha;\ |\alpha| < \varepsilon t]. \qquad (3.8)$$

If we use the inequalities:

$$\left| e^{i\alpha} - 1 - i\alpha + \alpha^2/2 \right| \leq \frac{1}{3}|\alpha|^3, \quad \left| e^{i\alpha} - 1 - i\alpha \right| \leq |\alpha|^2,$$

we obtain that

$$\left| \rho(k,n) \right| \leq \frac{|t|^3 b(k,n)}{6 B_n^2} \varepsilon + \frac{t^2}{B_n^2} \delta_{kn}(\varepsilon B_n);$$

and, consequently,

$$\left| \sum_{k=1}^{n} \nu(k,n)\rho(k,n) \right| \leq \varepsilon \frac{|t|^3}{6B_n^2} \tilde{S}_n^{(2)} + t^2 L_n(\varepsilon). \tag{3.9}$$

Since $M\tilde{S}_n^{(2)} \leq B_n^2$, it is easy to show, using Chebyshev's inequality, that the first term in (3.9) converges to zero in probability as $n \to \infty$ for any fixed t. Hence, if $L_n(\varepsilon) \xrightarrow{P} 0$ as $n \to \infty$ and for any $\varepsilon > 0$, then (3.3) is true according to (3.7).

Now we prove the necessity of the condition. Let relation (3.3) be true. Then we obtain from (3.7) that

$$\left(\exp\left\{ \sum_{k=1}^{n} \nu(k,n)\rho(k,n) \right\} - 1 \right) \psi_n(t) \xrightarrow{P} 0, \quad n \to \infty.$$

This means that as $n \to \infty$

$$P\left\{ \left| \exp\left\{ \sum_{k=1}^{n} \nu(k,n)\rho(k,n) \right\} - 1 \right| \exp\left\{ -\frac{t^2}{2B_n^2} \tilde{S}_n^{(2)} \right\} > \varepsilon \right\} \to 0 \tag{3.10}$$

for any $\varepsilon > 0$.

We have from the definitions of $\tilde{S}_n^{(2)}$ and B_n^2 that

$$P\left\{ \tilde{S}_n^{(2)} / B_n^2 > T \right\} \leq \frac{1}{T} \tag{3.11}$$

for all $n \in N$ and $T > 0$.

Choose $\varepsilon_1 > 0$ and define $R(n)$ by the relation

$$R(n) = \sum_{k=1}^{n} \nu(k,n)\rho(k,n).$$

Taking into account (3.11), by the total probability formula we have:

$$P\left\{ \left| e^{R(n)} - 1 \right| > \varepsilon_1 \right\} \leq P\left\{ \left| e^{R(n)} - 1 \right| e^{-\frac{t^2 \tilde{S}_n^{(2)}}{2B_n^2}} > \varepsilon_1 e^{-t^2 T} \right\} + \frac{1}{T}.$$

Let $\delta > 0$. First, we choose T such that $T^{-1} < \delta/2$. Then we fix this T and using (3.10) choose n such that

§1.3 Necessary and Sufficient Conditions

$$P\left\{\left|e^{R(n)}-1\right| e^{-\frac{t^2 \tilde{S}_n^{(2)}}{2B_n^2}} > \varepsilon_1 e^{-t^2 T}\right\} < \frac{\delta}{2}$$

for fixed ε_1, t and T. Hence,

$$P\left\{\left|e^{R(n)}-1\right| > \varepsilon_1\right\} < \delta$$

for a sufficiently large n. This means that

$$\sum_{k=1}^{n} \nu(k,n)\rho(k,n) \xrightarrow{P} 0, \quad n \to \infty. \qquad (3.12)$$

Further, we use standard arguments, developed in order to prove the necessity of the Lindeberg condition in the Central limit theorem. Namely, we use (see Borovkov, 1986) the fact that there is some $C(\varepsilon)>0$ such that the inequality

$$x^2 \leq \frac{1}{C(\varepsilon)} \operatorname{Re}(e^{ix} - 1 - ix + x^2/2) \qquad (3.13)$$

is true for any $\varepsilon>0$ and all $|x| \geq \varepsilon$.

If we put $x=(X(k,n)-a(k,n))t/B_n$ in (3.13), we have

$$\frac{t^2}{B_n^2} \delta_{kn}(\varepsilon B_n) \leq \frac{1}{C(\varepsilon)} \operatorname{Re} \rho(k,n)$$

for any $\varepsilon>0$ and fixed t. Therefore,

$$t^2 L_n(\varepsilon) \leq \frac{1}{C(\varepsilon)} \left|\sum_{k=1}^{n} \nu(k,n)\rho(k,n)\right|.$$

From the last inequality by (3.12) we obtain that $L_n(\varepsilon) \xrightarrow{P} 0$ as $n \to \infty$. The theorem is proved.

§3b. The Rate of Convergence

Assume that families of random variables $\{\nu(k,n)\}$ and $\{X(k,n)\}$ are independent. Let us introduce the following notations:

$$C_\delta(k,n) = M\left|X(k,n) - a(k,n)\right|^{2+\delta}, \quad 0<\delta\leq 1,$$

36   I. Sums of a Random Number of Random Variables

$$V_n^*(x) = V_n\left(\frac{B_n}{\delta_n}x\right), \quad \chi(n) = \sup_x \left|U_n(x) - \Phi_{\Delta_n/B_n} * V_n^*(x)\right|,$$

$$\gamma(k,n) = M\,|\nu(k,n)-\alpha(k,n)|.$$

Let $\Phi_a(x)$ be a normal distribution with zero expectation and variance a^2, and let C be some positive constants.

Theorem 3.2. Inequality

$$\chi(n) \le C\left[\frac{1}{\Delta_n^{2+\delta}}M\left[\sum_{k=1}^n \nu(k,n)C_\delta(k,n)\right] + M\left|\frac{\tilde{S}_n^{(2)}}{\Delta_n^2} - 1\right|\right] \qquad (3.14)$$

is true for some C and n∈N.

Proof. Let $T_o=\{0,1\}$. We denote by T_o^n the space of vectors $l=(l_1,\ldots,l_n)$ with coordinates 0 and 1, and put

$$S_n(l) = \sum_{k=1}^n l_k X(k,n), \qquad A_n(l) = \sum_{k=1}^n l_k a(k,n),$$

$$B_n^2(l) = \sum_{k=1}^n l_k b(k,n), \qquad P_n(l) = P\{\nu(k,n)=l_k,\ k=1,\ldots,n\}.$$

By the total probability formula we obtain that

$$\overline{U}_n(x) \stackrel{def}{=} \Phi_{\Delta_n/B_n} * V_n^*(x) = \sum_{l\in T_o^n} \Phi\left(\frac{Q_n(l,x)}{\Delta_n}\right) P_n(l),$$

where $Q_n(l,x) = B_n(x) - A_n(l) + A_n$. We represent the limiting distribution in the form:

$$U_n(x) = \sum_{l\in T_o^n} P\left\{\frac{S_n(1)-A_n(1)}{B_n(1)} < \frac{Q_n(1,x)}{B_n(1)}\right\} P_n(1).$$

Setting

$$\pi_n(1,x) = P\left\{\frac{S_n(1)-A_n(1)}{B_n(1)} < \frac{Q_n(1,x)}{B_n(1)}\right\} - \Phi\left(\frac{Q_n(1,x)}{\Delta_n}\right),$$

we obtain for any fixed n that

§1.3 Necessary and Sufficient Conditions

$$|U_n(x) - \bar{U}_n(x)| \le \sum_{1 \in \mathcal{K}_n} |\pi_n(1,x)| P_n(1) + \sum_{1 \in \mathcal{K}'_n} |\pi_n(1,x)| P_n(1), \quad (3.15)$$

where $\mathcal{K}_n$ is the set of all vectors $1 \in T_o^n$, for which

$$|B_n^2(1) - \Delta_n^2| \le \Delta_n^2/2, \quad \mathcal{K}'_n = \{1 \in T_o^n : |B_n^2(1) - \Delta_n^2| > \Delta_n^2/2\}.$$

If we use Chebyshev's inequality, we obtain that the second term in (3.15) is less than $4\Delta_n^{-2} M |\tilde{S}_n^{(2)} - \Delta_n^2|$. Represent the first term in the form:

$$\sum_{1 \in \mathcal{K}_n} |\pi_n(1,x)| P_n(1) \le I_1 + I_2, \quad (3.16)$$

where

$$I_1 = \sum_{1 \in \mathcal{K}_n} \left| P\left\{ \frac{S_n(1) - A_n(1)}{B_n(1)} < \frac{Q_n(1,x)}{B_n(1)} \right\} - \Phi\left(\frac{Q_n(1,x)}{B_n(1)} \right) \right| P_n(1),$$

$$I_2 = \sum_{1 \in \mathcal{K}_n} \left| \Phi\left(\frac{Q_n(1,x)}{B_n(1)} \right) - \Phi\left(\frac{Q_n(1,x)}{\Delta_n} \right) \right| P_n(1).$$

If $1 \in \mathcal{K}_n$, then

$$\frac{1}{2} \Delta_n^2 \le B_n^2(1) \le \frac{3}{2} \Delta_n^2. \quad (3.17)$$

In order to estimate I_1, we use Essen's inequality (Petrov, 1972, p.144) and have:

$$\sup_x I_1 \le C \sum_{1 \in \mathcal{K}_n} \frac{1}{B_n^{2+\delta}(1)} \sum_{k=1}^n 1_k C_\delta(k,n) P_n(1). \quad (3.18)$$

It follows from (3.17) and (3.18) that

$$\sup_x I_1 \le \frac{C}{\Delta_n^{2+\delta}} M\left[\sum_{k=1}^n \nu(k,n) C_\delta(k,n) \right]. \quad (3.19)$$

We use the following inequality (Petrov, 1972, p.143)

$$\sup_x |\Phi(px) - \Phi(x)| \sqrt{2\pi e} \le \begin{cases} p-1, & p \ge 1, \\ \frac{1}{p} - 1, & 0 < p < 1, \end{cases} \quad (3.20)$$

We obtain by means of (3.17) and (3.20) that

$$\sup_{x} I_2 \leq C \sum_{1 \in \mathcal{K}_n} \frac{1}{\Delta_n} |\Delta_n - B_n(1)| P_n(1).$$

Hence, using the inequality $|x-y|x^{-1} \leq |x^2-y^2|x^{-2}$, by the definition of $B_n(1)$ we have:

$$\sup_{x} I_2 \leq C\Delta_n^{-2} M|\tilde{S}_n^{(2)} - \Delta_n^2|. \qquad (3.21)$$

The statement of Theorem 3.2 follows from relations (3.15), (3.16), (3.19) and (3.21). Theorem 3.2 is thus proved.

Let $R_\delta(n)$ be the quantity in brackets on the right-hand side of (3.14), $r(n) = \sup_x |V_n(x)-V(x)|$. Denote by $\chi_1(n)$, $\chi_2(n)$ and $\chi_3(n)$ uniform deviations limit distributions and limiting ones in P.1°, P.2°, P.3° of Corollary 3.2.

The following statements establish uniform estimations in Corollary 3.2.

Corollary 3.3. If the conditions of the P.1° of Corollary 3.2 are satisfied, then there exists a C such that

$$\chi_1(n) \leq C\left[R_\delta(n) + \frac{1}{\Delta_n} M\left|\tilde{S}_n^{(1)} - A_n\right| + \sigma_n^2/\Delta_n^2\right]$$

for any $n=1, 2, \ldots$.

Corollary 3.4. Let the conditions of the P.2° of Corollary 3.2 be satisfied, and the function $V(x)$ be continuous. Then, for any $\varepsilon>0$,

$$\chi_2(n) \leq C\left[R_\delta(n)+r(n)+\delta_n(\varepsilon) + \frac{\Delta_n}{\sigma_n^2} M\left|\tilde{S}_n^{(1)} - A_n\right|\right],$$

where

$$\delta_n(\varepsilon) = \left(\frac{\Delta_n}{\varepsilon\sigma_n}\right)^3 + \sup_x |V(x+\varepsilon) - V(x-\varepsilon)|.$$

Corollary 3.5. Under the conditions of the P.3° of Corollary 3.2 there exists C such that for any $n=1,2,\ldots$

$$\chi_3(n) \leq C\left[R_\delta(n)+r(n) + \frac{1}{1-\theta}\left|\frac{\sigma_n^2}{B_n^2} - \theta\right|\left(1+\frac{B_n\sqrt{1-\theta}}{\sigma_n^2}M\left|\tilde{S}_n^{(1)} - A_n\right|\right)\right],$$

where

§1.3 Necessary and Sufficient Conditions

$$0 < \theta = \lim_{n \to \infty} \frac{\sigma_n^2}{B_n^2} < 1.$$

We will prove one of the corollaries, Corollary 3.3, for example. Since

$$\chi_1(n) \leq \chi(n) + \hat{\chi}_1(n), \qquad (3.22)$$

where

$$\hat{\chi}_1(n) = \sup_x \left| \Phi_{\Delta_n/B_n} * V_n^*(x) - \Phi(x) \right|,$$

and the estimate of $\chi(n)$ is known, it is sufficient to estimate the quantity $\hat{\chi}_1(n)$. If we use the total probability formula, then we have:

$$\hat{\chi}_1(n) \leq \sum_{l \in T_o^n} \sup_x \left| \Phi\left(\frac{Q_n(1,x)}{\Delta_n}\right) - \Phi(x) \right| P_n(1) \leq \sum_{l \in T_o^n} \sup_x \left| \Phi\left(\frac{B_n x}{\Delta_n}\right) - \Phi(x) \right| P_n(1)$$

$$+ \sum_{l \in T_o^n} \sup_x \left| \Phi\left(x - \frac{A_n(1)-A_n}{\Delta_n}\right) - \Phi(x) \right| P_n(1).$$

The first term of the last relation is less than $C\sigma_n^2/\Delta_n^2$, on the strength of (3.20). In order to estimate the second term, we use the inequality

$$\sup_x \left| \Phi(x+a) - \Phi(x) \right| \leq \frac{a}{\sqrt{2\pi}},$$

and obtain that it is not greater than

$$C\Delta_n^{-1} M |S_n^{(1)} - A_n|.$$

Therefore, for $n=1,2,\ldots,$

$$\hat{\chi}_1(n) \leq C \left[\frac{\sigma_n^2}{\Delta_n^2} + \frac{1}{\Delta_n} M \left| \tilde{S}_n^{(1)} - A_n \right| \right],$$

and this inequality, by (3.22) and by the estimates of quantity $\chi(n)$ from Theorem 3.2, proves Corollary 3.3.

Now we provide a non-uniform estimation in the case of the normal limit distribution.

Theorem 3.3. If $MX(k,n) = 0$, $M|X(k,n)|^{2+\delta} < \infty$, $0 < \delta \leq 1$, then there exists a constant C such that for any $n=1, 2, \ldots$

I. Sums of a Random Number of Random Variables

$$|U_n(x)-\Phi(x)| \leq \frac{C}{(1+|x|)^{2+\delta}}\left[R_\delta(n) + \frac{1}{\Delta_n^4} M\left|\tilde{S}_n^{(2)} - \Delta_n^2\right|^2\right].$$

Proof. It is clear that there exists an absolute constant A such that

$$(1+|x|^3)\left(\frac{x^2}{12 \ln(2+|x|)} - 1\right)^{-2} < 1 \qquad (3.23)$$

for $|x|>A$. Let $|x|>A$. On the strength of the total probability formula

$$U_n(x) = \sum_{l \in T_o^n} P\left\{\frac{S_n(1)}{B_n(1)} < \frac{\Delta_n x}{B_n(1)}\right\} P_n(1).$$

Therefore the following estimation is true:

$$|U_n(x)-\Phi(x)| \leq \sum_{l \in T_o^n} \left|P\left\{\frac{S_n(1)}{B_n(1)} < \frac{\Delta_n x}{B_n(1)}\right\} - \Phi\left(\frac{\Delta_n x}{B_n(1)}\right)\right| P_n(1)$$

$$+ \sum_{l \in T_o^n} \left|\Phi\left(\frac{\Delta_n x}{B_n(1)}\right) - \Phi(x)\right| P_n(1). \qquad (3.24)$$

In order to estimate the first sum, we use the following version of Bikelis's (1966) theorem.

Theorem A. If $MX(k,n)=0$, $M|X(k,n)|^{2+\delta}<\infty$, $0<\delta\leq 1$, then

$$\left|P\left\{\frac{S_n(1)}{B_n(1)} < x\right\} - \Phi(x)\right| \leq \frac{C}{B_n^{2+\delta}(1)(1+|x|)^{2+\delta}} \sum_{k=1}^n 1_k M|X(k,n)|^{2+\delta}.$$

Represent the set of summation in the first term of (3.24) in the form:

$$T_o^n = \mathcal{R}_n \cup \mathcal{R}'_n, \qquad (3.25)$$

where

$$\mathcal{R}_n = \{l \in T_o^n : B_n(1) \geq \Delta_n\}, \quad \mathcal{R}'_n = \{l \in T_o^n : B_n(1) < \Delta_n\}.$$

Then, if we use Theorem A, we obtain that the sum with respect to l from the set $\mathcal{R}_n$ is less than

§1.3 Necessary and Sufficient Conditions

$$\frac{C}{\Delta_n^{2+\delta}} \sum_{l \in \mathcal{R}_n} \frac{1}{\left(\frac{B_n(1)}{\Delta_n} + |x|\right)^{2+\delta}} \sum_{k=1}^{n} l_k c_\delta(k,n) P_n(1)$$

$$\leq \frac{C}{\Delta_n^{2+\delta}(1+|x|)^{2+\delta}} M\left[\sum_{k=1}^{n} \nu(k,n) c_\delta(k,n)\right].$$

Since $A>1$, in order to estimate the sum with respect to l from $\mathcal{R}_n'$, we can use the inequality

$$(1+|x|)/(a+|x|)<2, \quad |x|>1, \quad a<1.$$

Then, on the strength of Theorem A, it is less than

$$\frac{C}{\Delta_n^{2+\delta}(1+|x|)^{2+\delta}} M\left[\sum_{k=1}^{n} \nu(k,n) c_\delta(k,n)\right].$$

Now we estimate the second term in (3.24). Use Prokhorov's (1960) lemma, which can be written in the form:

$$\left|\Phi(x/\sigma_1) - \Phi(x/\sigma_2)\right| \leq \frac{1}{\sqrt{\pi}} \exp\left\{-x^2/4\sigma_2^2\right\} \left|\sigma_1/\sigma_2 - 1\right|. \tag{3.26}$$

If we use (3.26) in the second term of (3.24), we obtain the following estimation for the sum with respect to l from $\mathcal{R}_n$:

$$\frac{1}{\Delta_n \sqrt{\pi}} e^{-x^4/4} \sum_{l \in \mathcal{R}_n} |B_n(1) - \Delta_n| P_n(1)$$

$$\leq \frac{C}{\Delta_n^2 (1+|x|)^{2+\delta}} M \left|\tilde{S}_n^{(2)} - \Delta_n^2\right|. \tag{3.27}$$

Simple calculations show that $4(2+\delta)x^{-2}\ln(2+|x|)<1$ for $|x|>A$. Therefore, setting

$$\varepsilon^2 = x^2 \Delta_n^2 / 4(2+\delta)\ln(2+|x|),$$

we can represent the set $\mathcal{R}_n'$ from (3.25) in the form:

$$\mathcal{R}_n' = T_1(n,\varepsilon) \cup T_2(n,\varepsilon),$$

where

$$T_1(n,\varepsilon) = \{l \in T_o^n : \Delta_n \leq B_n(1) \leq \varepsilon\}, \quad T_2(n,\varepsilon) = \{l \in T_o^n : B_n(1) > \varepsilon\}.$$

The sum with respect to l from $T_1(n,\varepsilon)$ has the estimate

42 I. Sums of a Random Number of Random Variables

$$\frac{1}{\Delta_n^2 \sqrt{\pi}} \exp\{-x^2\Delta_n^2 / 4\varepsilon^2\} M \left|\tilde{S}_n^{(2)} - \Delta_n^2\right|.$$

This quantity is not greater than the expression on the right-hand side of (3.27). Finally, the sum with respect to l from the set $T_2(n,\varepsilon)$ is bounded by

$$\sum_{l \in T_2(n,\varepsilon)} \left| P\left\{\frac{S_n(l)}{B_n(l)} < \frac{\Delta_n x}{B_n(l)}\right\} - \Phi\left(\frac{\Delta_n x}{B_n(l)}\right) \right| \leq |\varepsilon^2 - \Delta_n^2|^{-2} \sum_{l \in T_2(n,\varepsilon)} |B_n^2(l) - \Delta_n^2| P_n(l),$$

where $(\varepsilon^2 - \Delta_n^2)^2 \geq \Delta_n^4 (1+|x|)^{2+\delta}$. If we use definitions of quantities Δ_n^2 and $B_n^2(l)$, we have:

$$\sum_{l \in T_2(n,\varepsilon)} |B_n^2(l) - \Delta_n^2|^2 P_n(l) \leq \beta^2(n).$$

Thus, the sum with respect to l from $T_2(n,\varepsilon)$ is not greater than $\beta^2(n)/\Delta_n^4 (1+|x|)^{2+\delta}$ and the theorem is proved for $|x|>A$. When $|x|\leq A$ the statement of the theorem follows from uniform estimation. The theorem is thus proved.

Let $X(i,N)$, $i=1,\ldots,N$ be independent random variables, $\nu(i,N)$, $i=1,\ldots,N$, be random variables taking values 0 and 1. Suppose that vectors $(X(1,N),\ldots,X(N,N))$ and $(\nu(1,N),\ldots,\nu(N,N))$ are independent for any N. Let $\{T_i, i\in N\}$ be a sequence of random variables such that the variable T_i takes values $1,2,\ldots,i$. Consider the sampling sum V_N of a random sample size T_N. Then, under the notations introduced above,

$$M[\nu(k,n)|T_N=1] = 1/N, \quad M\nu(k,N) = MT_N/N \equiv f_N,$$

$$\sigma_N^2 = D\tilde{S}_N^{(1)} = \left[\frac{N}{N-1} f_N(1-f_N) - \frac{DT_N}{N(N-1)}\right]\alpha(N),$$

$$\Delta_N^2 = f_N\beta(N), \quad \alpha(N) = \sum_{k=1}^{N} a^2(k,N), \quad \beta(N) = \sum_{k=1}^{N} b(k,N).$$

Suppose that the following limit exists:

$$\lim_{N\to\infty} \sigma_N^2 / \Delta_N^2 = C \in [0,\infty]. \qquad (3.28)$$

Denote $\hat{X}(k,n) = X(k,n) - a(k,n)$,

§1.3 Necessary and Sufficient Conditions

$$P_N^{(1)}(x) = P\left\{\frac{V_N}{B_N} < x\right\}, \quad P_N^{(2)} = P\left\{\frac{\tilde{S}_N^{(1)}}{\sigma_N} < x\right\}$$

and introduce the conditions:

$$\max_{1 \leq k \leq N} b(k,N) = o\left(\sum_{k=1}^{N} b(k,N)\right), \quad N \to \infty, \tag{3.29}$$

$$[\beta(N)]^{-1} \sum_{k=1}^{N} M\left[\hat{X}^2(k,N); \frac{|\hat{X}(k,N)|}{\beta^2(N)} > \varepsilon\right] \to 0. \tag{3.30}$$

Corollary 3.6. Let condition (3.28) be satisfied. Then:

$1°$ If $C=0$, $f_N \to 1$ and (3.29) and (3.30) are satisfied, then

$$\lim_{N \to \infty} P_N^{(1)}(x) = \Phi(x);$$

$2°$ if $C=\infty$, then

$$\lim_{N \to \infty} P_N^{(1)}(x) = \lim_{N \to \infty} P_N^{(2)}(x);$$

$3°$ if $C \in (0,\infty)$, $f_N \to 1$, conditions (3.29) and (3.30) are satisfied and $P_N^{(2)}(x) \to P^{(2)}(x)$, as $N \to \infty$, then

$$\lim_{N \to \infty} P_N^{(1)}(x) = \int_{-\infty}^{\infty} \Phi((x-u)\sqrt{1+C}) dP^{(2)}\left(u\sqrt{\frac{1+C}{C}}\right).$$

Now we will mention an estimate of the distance between the limit and limiting distributions. Let

$$C(k,N) = M|X(k,N)-a(k,N)|^3, \quad \chi(N) = \sup_{x} |P_N^{(1)}(x)-\Phi(x)|.$$

Corollary 3.7. If all $C(k,N)<\infty$, then

$$\chi(N) \leq C\left[\frac{\max\limits_{1 \leq k \leq N} C(k,N)}{\sqrt{MT_N}\left(\frac{1}{N}\beta(N)\right)^{3/2}} + (1-f_N)(1+\alpha(N)/\beta(N))\right]$$

for some absolute constant C and $N \geq 1$.

CHAPTER II

BRANCHING PROCESSES WITH GENERALIZED IMMIGRATION

§2.1. CLASSICAL MODELS OF BRANCHING PROCESSES

We will use the following intuitive description of branching processes instead of the formal definition of the process that corresponds to its real substance in connection with physical and chemical emergences. Consider the scheme of evolution and reproduction of some particles. Each of these particles, independently of the others, lives in a random time L and generates a random number ν of new particles. These new particles undergo analogous transformations.

Models of branching processes differ from each other by additional assumptions on the reproduction process of new particles and on the distribution of the vector (L,ν). In the theory of branching processes the main object of investigation is a random process $\mu(t)$ (size of population at time $t \in [0,\infty)$).

Let each particle have a life-length of one time unit (L=1) and assume that at its death it splits into ν new particles. In this case the process $\mu(t)$ is called a Galton-Watson process. If random variables L and ν are independent and $P\{L \le x\} = 1 - e^{-\lambda x}$, $x>0$, $\lambda>0$, then process $\mu(t)$ is called a continuous time Markov branching process.

A natural generalization of the continuous time Markov branching process is to allow lifetimes to be random variables with arbitrary distribution $G(t)$. Besides, we assume that L and ν are independent, the pairs (L,ν) are independent and identically distributed for different particles and that a realized particle is born at the death of its mother. In this case the process $\mu(t)$ is called a Bellman-Harris process. In contrast to Galton-Watson and continuous time branching processes, this process is not Markovian.

If in the Bellman-Harris process we allow the dependence of ν on the lifetime L, we have a real age-dependent process. This process is called a Sevast'yanov model.

§2.1. Classical Models of Branching Processes

Let us assume that particles beget children at randomly chosen instants during their lives and not necessarily exactly when they die. We also allow that ν depends on L, and new particles generate according to some point process $\xi(a)$ and, as before, the pairs (L,ξ) for different particles are independent and identically distributed except, possibly, for different distributions for ancestors. Note that the $\xi(a)$ means the number of births up to and including age t. In this case $\mu(t)$ is called a general Kramp-Mod-Jagers process.

For details about general models of branching processes see, for example, Sevast'yanov (1971) and Jagers (1975). The basic purpose of this section is:
a) to demonstrate methods of the theory of branching processes by proving some limit theorems for simple models;
b) to give some known results which will be used in the following sections.

§1a. Bellman-Harris Processes

We start from some statements about Bellman-Harris processes. Let the process begin at time t=0 from an ancestor having a random lifetime τ_0 and which at its death splits into k new particles with probability P_k. Here $\{P_k, k \in N_0\}$ is some probability distribution. These new particles have random lifetimes $\tau_{11}, \tau_{12}, \ldots, \tau_{1k}$, respectively and at their deaths they reproduce according to the law $\{P_k\}$. Suppose that lifetimes τ_0, τ_{ij}, i, j=1,2,... are independent and identically distributed positive random variables with distribution G(t). Suppose also the independence of lifetimes of the particles and the number of their descendants. As we note above, in this case $\mu(t)$, $\mu(0)=1$, is a one-dimensional Bellman-Harris process. In view of our assumptions, the investigation of $\mu(t)$ can be reduced to the analysis of the following equation for generating functions:

$$F(t,s) = s[1-G(t)] + \int_0^t f(F(t-u,s))dG(u), \quad |s|\leq 1, \qquad (1.1)$$

where

$$F(t,s) = \sum_{k=0}^{\infty} P\{\mu(t)=k\}s^k, \quad f(s) = \sum_{k=0}^{\infty} p_k s^k.$$

The Bellman-Harris process includes simpler models of Galton-Watson processes and continuous time Markov branching processes. In fact, if G(t) is the step function:

then (1.1) becomes part of the iterative formula

$$F(t,s) = f(F(t-1,s)), \quad F(1,s) = f(s), \quad (1.2)$$

defining the generating function of the Galton-Watson process. But if $G(t)$ is the exponential distribution:

$$G(t)g = \begin{cases} 0, & t<0 \\ 1-e^{-\lambda t}, & t\geq 0 \end{cases},$$

then (1.1) becomes part of the differential equation

$$\frac{F(t,s)}{t} = \lambda[f(F(t,s))-F(t,s)], \quad F(0,s)=s, \quad (1.3)$$

and the solution of this equation is a generating function of a continuous time Markov branching process.

Now we consider continuous and discrete time Markov models of the branching process in detail. Transitional probabilities $P_{ij}(t)=P\{\mu(t)=j \mid \mu(0)=i\}$ of these processes satisfy the Markov property condition:

$$P_{ij}(t) = \sum_{k=0}^{\infty} P_{ik}(\tau)P_{kj}(t-\tau) \quad (1.4)$$

for any $i, j \in \mathbb{N}$, $\tau, t \in [0,\infty)$, $0 \leq \tau \leq t$.

§1b. Moments and Extinction Probabilities

1 Moments. Factorial moments of the process

$$M_k(t) = M\mu(t)(\mu(t)-1)\ldots(\mu(t)-k+1)$$

are defined by the following formula:

$$M_k(t) = \left.\frac{\partial^k F(t,s)}{\partial s^k}\right|_{s=1}. \quad (1.5)$$

Let us denote $A=M_1(1)$, $B=M_2(1)$ for Galton-Watson processes and $a=f'_*(1)$, $b=f''_*(1)$ for continuous time processes, where $f_*(s) = \lambda(f(s)-s)$.

If we use formulas (1.2) and (1.5), we obtain:

§2.1. Classical Models of Branching Processes

$$M\mu(t) = A^t,$$

$$D\mu(t) = \begin{cases} \dfrac{B+A-A^2}{A(A-1)} A^t(A^t-1), & A \neq 1, \\ Bt, & A = 1 \end{cases}$$

for Galton-Watson processes. For the continuous time process we use the equation (1.3) and the formula (1.5) and obtain:

$$M\mu(t) = e^{at},$$

$$D\mu(t) = \begin{cases} (\dfrac{b}{a}-1)e^{at}(e^{at}-1), & a \neq 0, \\ bt, & a = 0 \end{cases}$$

2 The Extinction Probability.

The formulas obtained above show that the asymptotic behavior of the expectation and variance of the process substantially depends on quantities A and a. Due to this we divide the set of all branching processes into subcritical, critical and supercritical classes.

Definition 1.1. We shall call the discrete time branching process subcritical, critical or supercritical, according to whether $A<1$, or $A=1$ and $B>0$, or $A>1$ respectively.

Definition 1.2. We shall call the continuous time branching process subcritical, critical or supercritical, according to whether $a<0$, or $a=0$, and $b>0$, or $a>0$ respectively.

Definition 1.3. The probability of the event $\{\mu(t)=0$ for some $t\}$ is called an extinction probability of the process $\mu(t)$. The process is called degenerate, if this probability is equal to 1.

Theorem 1.1. The extinction probability q of the process $\mu(t)$ is equal to the following limit:

$$q = \lim_{t \to \infty} P\{\mu(t)=0\}. \qquad (1.6)$$

Proof. The existence of the limit (1.6) follows from the inequality

$$P\{\mu(t)=0\} \leq P\{\mu(\tau)=0\}$$

for all $\tau \geq t$. If we denote by A an event consisting of the degeneration of the process $\mu(t)$, then

$$A = \bigcup_{k=1}^{\infty} \{\mu(k)=0\}.$$

Since

$$\lim_{t \to \infty} P\{\mu(t)=0\} = \lim_{n \to \infty} P\{\mu(n)=0\},$$

using the relation

$$P(A) = \lim_{n \to \infty} P\{\mu(n)=0\}$$

we obtain (1.6). The theorem is proved.

A reader can find the proof of the following statement, for example, in the books of Sevast'yanov (1971) or of Athrea and Ney (1972).

Theorem 1.2. The extinction probability q of the Bellman-Harris process, also of the Markov branching process in particular, is equal to the smallest non-negative solution of the equation $f(s)=s$.

It follows from Theorem 1.2 that $q=1$ if and only if $f'(1) \leq 1$, that is the process is degenerate if and only if it is subcritical or critical. Since subcritical and critical processes degenerate with probability 1, $P\{\mu(t)>0\} \to 0$ as $t \to \infty$ for such processes. The quantity $Q(t) = P\{\mu(t)>0\}$ is called the probability of non-extinction of the process. The form of limit theorems substantially depends on the rate of convergence of $Q(t)$ to zero. We demonstrate this fact for critical Galton-Watson process by obtaining an estimate of the rate of convergence of $Q(t)$.

§1c. Asymptotics of the Probability of Non-Extinction and the Exponential Limit Distribution

Theorem 1.3. If in the critical Galton-Watson process $B \in (0, \infty)$, then

$$\lim_{t \to \infty} \frac{Bt}{2} Q(t) = 1.$$

We mention the following property of probability generating functions from Sevast'yanov's book (1971).

Lemma 1.1. If the probability generating function $F(s)$ has a kth order derivative $F^{(k)}(1)$ at the point $s=1$, then the expansion

$$F(s) = \sum_{i=1}^{k-1} F^{(i)}(1) \frac{(s-1)^i}{i!} + R_k(s) \frac{(s-1)^k}{k!}$$

is true. If s is real-valued, then

§2.1. Classical Models of Branching Processes

$$0 \leq R_k(s) \leq F^{(k)}(1), \quad 0 \leq s \leq 1,$$

and is non-decreasing with respect to s. If s is complex-valued, then

$$|R_k(s)| \leq F^{(k)}(1), \quad |s| \leq 1,$$

and $R_k(s) \to F^{(k)}(1)$ as $s \to 1$.

Proof of Theorem 1.3. We obtain from relation (1.2) that

$$Q(t) = 1 - f(1 - Q(t-1)).$$

Expanding $f(s)$ with the help of Lemma 1.1 we have:

$$Q(t) = Q(t-1) - \frac{B + \varepsilon_1(t)}{2} Q^2(t-1), \qquad (1.7)$$

where $\varepsilon_1(t) \to 0$ as $t \to \infty$. We can rewrite (1.7) in the form:

$$Q(t) = Q(t-1) - \frac{B}{2} Q(t-1)Q(t) + \varepsilon_2(t), \qquad (1.8)$$

where

$$\varepsilon_2(t) = - \frac{\varepsilon_1(t)}{2} Q^2(t-1) - \frac{B}{2} Q^3(t-1) \frac{B + \varepsilon_1(t)}{2}.$$

Since it follows from (1.7) that fraction $Q(t)/Q(t-1) \to 1$ as $t \to \infty$, then

$$\lim_{t \to \infty} [\varepsilon_2(t)/Q(t)Q(t-1)] = 0. \qquad (1.9)$$

Thus, in the relation

$$\frac{1}{Q(t)} = \frac{1}{Q(t-1)} + \frac{B}{2} + \varepsilon_3(t), \qquad (1.10)$$

obtained from (1.8) (if we divide by $Q(t)Q(t-1)$), the quantity $\varepsilon_3(t)$ tends to 0 as $t \to \infty$. The sequential application of (1.10) gives the following equality:

$$\frac{1}{Q(t)} = 1 + Bt/2 + \sum_{k=0}^{t-1} \varepsilon_3(k).$$

The theorem is thus proved.

Theorem 1.4. Under the conditions of Theorem 1.3

$$\lim_{t \to \infty} P\left\{ \frac{2\mu(t)}{Bt} \geq x \mid \mu(t) > 0 \right\} = e^{-x}, \quad x \geq 0.$$

Theorem 1.4 is one of the well-known classic results of the theory of branching processes. A reader can see different proofs of this theorem in the books of Sevast'yanov (1971), Athreya and Ney (1972), and of Jagers (1975). For example, in Athreya and Ney (1972) and Jagers (1975) the proof is based on the following statement:

II. Branching Processes with Generalized Immigration

Basic lemma. [Kesten, Ney, Spitzer (1966)]. If in the critical Galton - Watson process $B \in (0, \infty)$, then

$$\lim_{t \to \infty} \frac{1}{t} \left[\frac{1}{1-F(t,s)} - \frac{1}{1-s} \right] = \frac{B}{2} \qquad (1.11)$$

uniformly for $0 \le s < 1$, where $F(t,s) = Ms^{\mu(t)}$.

The proof is based on the following asymptotic formula in Sevast'yanov's book. Under conditions the of Theorem 1.4

$$1 - F(t,s) = \frac{1-s}{1 + \frac{Bt}{2}(1-s)} (1 + \varepsilon(t,s)), \qquad (1.12)$$

where $\varepsilon(t,s) \to 0$ as $t \to \infty$ uniformly for $|s| \le 1$. We will give another proof of this theorem suggested by K. Borovkov (1988).

Proof of Theorem 1.4. Let us denote by $\Phi_t(\lambda)$ the Laplace transform of $\{Q(t)\mu(t) | \mu(t) > 0\}$. Then

$$\Phi_t(\lambda) \sim M\left[e^{-\lambda B(t)\mu(t)} \Big| \mu(t) > 0 \right] \qquad (1.13)$$

as $t \to \infty$, for any function $B(t)$ such that $B(t) \sim Q(t)$. If we define $B(t)$ by equality $1 - e^{-\lambda B(t)} = \lambda Q(t)$, we obtain:

$$\Phi_t(\lambda) \sim \frac{1}{Q(t)} M\left[(1-\lambda Q(t))^{\mu(t)}; \mu(t) > 0 \right] \stackrel{\text{def}}{\equiv} f(t, \lambda). \qquad (1.14)$$

Introduce the function $H(u) = \inf\{t : Q(t) \le u\}$ for $0 < u < 1$. It is obvious that

$$Q(H(\lambda Q(t))) \le \lambda Q(t) \le Q(H(\lambda Q(t)) - \delta) \qquad (1.15)$$

for any $\delta > 0$. Since on the strength of the total probability formula

$$M\left[(1-Q(\tau))^{\mu(t)}; \mu(t) > 0 \right] = Q(t) - Q(t+\tau),$$

using (1.15) we obtain from (1.14) that

$$1 - q(t, \lambda) \frac{Q(t_1 - \delta)}{Q(t_1)} \le f(t, \lambda) \le 1 - q(t, \lambda), \qquad (1.16)$$

where $q(t, \lambda) = Q(t_1)/Q(t)$, $t_1 = t + H(\lambda Q(t))$.

Now we consider the function $q(t, \lambda)$. It follows from Theorem 1.3 that the function $Q(t)$ is representable in the form $Q(t) = t^{-1} M(t)$, where $M(t) \to 2/B$ as $t \to \infty$. Therefore, $H(u)$ can be written as $H(u) = u^{-1} M_1(u)$, where $M_1(u) \to 2/B$ as

§2.1. Classical Models of Branching Processes

$u \to 0$. Taking into account these facts and Theorem 1.3 we have:
$$Q(t_1) \sim Q((1+\lambda^{-1})t) \sim (1+\lambda^{-1})^{-1}Q(t),$$
that is, $q(t,\lambda) \to (1+\lambda^{-1})^{-1}$ as $t \to \infty$. Therefore, it follows from (1.14) and (1.16) that $\Phi_t(\lambda)$ converges to the Laplace transform of the exponential distribution. The theorem is thus proved.

In fact the method of proof, being carried out, is applicable in more general situations. For details we refer the reader to Borovkov (1988), Vatutin and Sagitov (1988, 1989).

§1d. Branching Processes with Stationary Immigration

Let $\{X_{it}, i \in N, t \in N_o\}$ and $\{\xi_t, t \in N_o\}$ be families of integer-valued non-negative random variables. Define the process $Z(t)$, $t \in N_o$, by
$$Z(0)=0, \quad Z(t+1) = \sum_{i=1}^{Z(t)} X_{it} + \xi_t, \qquad (1.17)$$
where $\sum_{i=1}^{o} = 0$. If we interpret random variables X_{it} and ξ_t as the number of direct descendants of ith particle at time t and as the number of immigrants at the time t respectively, then $Z(t)$ is the number of particles at the time t in the Galton-Watson process with immigration. Assume that the random variables ξ_t have the same distribution for different $t \in N_o$ and put
$$h(s) = \sum_{k=0}^{\infty} P\{\xi_t = k\} s^k.$$
Besides this, the random variables X_{it} have the same distribution for all pairs (i,t) and, what is more,
$$P\{X_{it}=k\}=p_k, \quad f(s) = \sum_{k=0}^{\infty} p_k s^k = M s^{X_{it}}.$$

According to the accepted terminology, in this case the "immigration process" is the sum of independent and identically distributed random variables
$$X(t) = \sum_{i=0}^{t} \xi_i.$$
If we denote by $\mu(t)$ the process without immigration, then
$$M\left[s^{\mu(t)} \Big| \mu(0)=1\right] = f_t(s),$$

where $f_t(s)$ is tth iteration of $f(s)$.

We are interested in the asymptotic behavior of $Z(t)$. If we denote $H(t,s)=MS^{Z(t)}$, then we obtain from (1.17)

$$H(t+1,s) = H(t,f(s))h(s). \qquad (1.18)$$

Consequently,

$$H(t,s) = \prod_{k=0}^{t-1} h(f_k(s)). \qquad (1.19)$$

Theorem 1.5. If $A=f'(1)=1$, $B = f''(1) \in (0,\infty)$ and $\alpha = h'(1) = M\xi_t \in (0,\infty)$, then $2Z(t)/Bt$ as $t\to\infty$ converges in distribution to a random variable with the density function

$$\frac{1}{\Gamma(2\alpha/B)} x^{2\alpha/B-1} e^{-x}, \quad x>0.$$

Proof. First we consider the following sum:

$$\Sigma = \sum_{k=0}^{t-1} (1-h(f_k(s_t))), \quad s_t = e^{-2\tau/Bt}, \quad \tau>0.$$

The generating function $h(s)$ is representable in the form

$$1-h(s) = \alpha(1-s) - \varepsilon(s)(1-s) \qquad (1.20)$$

for any $s\in[0,1]$, where $\varepsilon(s)\to 0$ as $s\to 1$. If we use (1.20), it is possible to decompose the sum Σ as:

$$\Sigma = \alpha \sum_{k=0}^{t-1} (1-f_k(s_t)) + \sum_{k=0}^{t-1} \varepsilon(f_k(s_t))(1-f_k(s_t)). \qquad (1.21)$$

Let $N(t)$ be an integer-valued positive function, such that $N(t)\to\infty$ and $N(t)=o(t)$ as $t\to\infty$. Using relation (1.12) we have:

$$\lim_{t\to\infty} \sum_{k=N(t)+1}^{t-1} (1-f_k(s_t)) = 2B^{-1}\ln(1+\tau),$$

and the sum from 0 to $N(t)$ converges to zero as $t\to\infty$. Consequently, the first sum in (1.21) converges to $2\alpha B^{-1}\ln(1+\tau)$ as $t\to\infty$.

Since $\varepsilon(s)\to 0$ and for any $s\in[0,1]$ the function $f_k(s)$ is non-decreasing according to k in the critical case (see Sevast'yanov, 1971, p. 50), then the second sum in (1.21) is less than

$$\varepsilon(f_0(s_t)) \sum_{k=0}^{t-1} (1-f_k(s_t)).$$

This fact, with regard to the asymptotic behavior of the first part, shows

§2.1. Classical Models of Branching Processes   53

that the second sum in (1.21) converges to zero as $t \to \infty$. Thus,

$$\lim_{t \to \infty} \sum_{k=0}^{t-1} (1-h(f_k(s_t))) = \frac{2\alpha}{B} \ln(1+\tau).$$

If we use the elementary property of logarithmic function, we have:

$$\lim_{t \to \infty} \sum_{k=0}^{t-1} \ln h(f_k(s_t)) = - \lim_{t \to \infty} \sum_{k=0}^{t-1} (1-h(f_k(s_t))),$$

that is

$$\lim_{t \to \infty} \ln H(t, s_t) = - \frac{2\alpha}{B} \ln(1+\tau).$$

Hence Theorem 1.5 is proved.

The following theorem shows that there is a stationary limit distribution for $Z(t)$ in the subcritical case.

Theorem 1.6. a) If $A>1$ or $A=1$, $B \in (0,\infty)$, then

$$\lim_{t \to \infty} P\{Z(t)=k\} = 0$$

for any $k \in N$;

b) if $A<1$ and $\alpha \in (0,\infty)$, then

$$\lim_{t \to \infty} P\{Z(t)=k\} = \rho_k$$

exists for any $k \in N_0$ and $\{\rho_k, k \in N_0\}$ is a probability distribution.

Proof. Since $H(t,s)$ is decreasing according to t for any fixed $s \in [0,1)$, there exists the limit

$$H(s) = \lim_{t \to \infty} H(t,s) = \prod_{k=0}^{\infty} h(f_k(s)). \qquad (1.22)$$

Besides this, $H(s)>0$ if and only if the row

$$\sum_{k=0}^{\infty} (1-h(f_k(s)))$$

converges.

If $\alpha<\infty$, it follows from (1.20) that

$$1-h(f_k(s)) = \alpha(1-f_k(s)) + o(1-f_k(s)) \qquad (1.23)$$

as $k \to \infty$. Therefore, $H(s)>0$ if and only if

$$\sum_{k=0}^{\infty} (1-f_k(s)) = \infty. \qquad (1.24)$$

On the other hand, the row (1.24) diverges under the conditions of Part a and converges under the conditions of Part b.

If $\alpha=\infty$ and $A>1$, then $f_k(s) \to q$ as $k \to \infty$, where q is the extinction probability, and, consequently, $\sum_{k=0}^{\infty}(1-h(f_k(s)))=\infty$. If $\alpha=\infty$ and $A=1$, then

$$1-h(f_k(s)) \geq C(1-f_k(s))$$

for some C and, therefore,

$$\sum_{k=0}^{\infty}(1-h(f_k(s))) = \infty.$$

Further, under the conditions of Part b,

$$\lim_{s \to 1} H(s) = \lim_{t \to \infty} H(f_t(0)) = \lim_{t \to \infty} \prod_{k=0}^{\infty} h(f_{k+t}(0)) = 1.$$

The theorem is proved.

The following result gives a limit distribution in the supercritical case.

Theorem 1.7. If $A>1$, then there is a sequence of constants $\{C_t\}$ such that $\{Z(t)/C_t\}$ converges with probability 1 to a random variable Z. If $E\ln X_{11} < \infty$, then $P\{Z<\infty\}=1$ and Z has an absolutely continuous distribution on $(0,\infty)$. If $E\ln X_{11}=\infty$, then $P\{Z<\infty\}=0$.

Theorem 1.7 was proved by E. Seneta (1970).

§1e. Continuous Time Branching Processes with Immigration

Now we consider the continuous-time Markov branching process with homogeneous immigration. Let us consider some particles of the same type which act independently of each other. Each of these particles splits into k new particles during the time interval $(t,t+\Delta t)$ with the probability $\delta_{k1} + p_k(t)\Delta t + o(\Delta t)$, $\Delta t \to 0$. Besides this, k particles immigrate into the process during the time interval $(t,t+\Delta t)$ with the probability $\delta_{k0} + q_k(t)\Delta(t) + o(\Delta t)$. Here $\delta_{ij}=0$ if $i \neq j$ and $\delta_{ii}=1$.

It is obvious that quantities $p_k(t)$ and $q_k(t)$ must satisfy the following conditions for any $t \in [0,\infty)$:

§2.1. Classical Models of Branching Processes

$$p_1(t)<0, \; p_k(t)\geq 0, \; k\neq 1, \; \sum_{k=0}^{\infty} p_k(t)=0,$$
$$q_0(t)<0, \; q_k(t)\geq 0, \; k>0, \; \sum_{k=0}^{\infty} q_k(t)=0. \quad (1.25)$$

The assumption of homogeneity means that quantities $p_k(t)$ and $q_k(t)$ are the same for all t. Therefore, in this case, their generating functions are the same for all t, that is,

$$f_*(s)=\sum_{k=0}^{\infty} p_k(t)s^k = \sum_{k=0}^{\infty} p_k s^k, \quad h_*(s)=\sum_{k=0}^{\infty} q_k(t)s^k = \sum_{k=0}^{\infty} q_k s^k. \quad (1.26)$$

It is clear that in this case the immigration process X(t) is a compound Poisson process.

As before, we denote by $\mu(t)$ the number of particles at time t in the process without immigration and by Z(t) those in the process with immigration. We also put

$$F(t,s) = E\left[s^{\mu(t)} \Big| \mu(0)=1\right], \; H(t,s)=E\left[s^{Z(t)}\Big|Z(0)=0\right].$$

It is known from (1.3) that F(t,s) must satisfy the equation

$$\frac{F(t,s)}{t} = -p_1[f(F(t,s))-F(t,s)], \quad (1.27)$$

where

$$f(s) = \frac{f_*(s)-p_1 s}{-p_1}.$$

Besides this,

$$H(t,s) = \exp\left\{\int_0^t h_*(F(u,s))du\right\}. \quad (1.28)$$

Introduce the following notations for derivatives of functions $f_*(s)$ and $h_*(s)$:

$$f'_*(1)=a, \; f''_*(1)=b, \; h'_*(1)=\alpha_*, \; h''_*(1)=\beta_*.$$

If we use (1.28), it is possible to calculate the expectation and variance of Z(t) and to obtain the following asymptotes.

If a<0, then

$$\lim_{t\to\infty} EZ(t) = -\frac{\alpha_*}{a}, \quad \lim_{t\to\infty} DZ(t) = -\frac{\alpha_*}{a} + \frac{\alpha_* b - \beta_* a}{2a^2};$$

if a>0, then as $t\to\infty$

56 II. Branching Processes with Generalized Immigration

$$EZ(t) \sim \frac{\alpha_*}{a} e^{at}, \quad DZ(t) \sim \frac{\beta_* a - \alpha_* b}{2a^2} e^{2at};$$

but if a=0, then

$$EZ(t) = \alpha_* t, \quad DZ(t) = (\beta_* + \alpha_*)t + \frac{\alpha_* b_*}{2} t^2.$$

If a<0, there is a stationary limit distribution for $Z(t)$, as in the case of the discrete time process.

Theorem 1.8. If $a<0$, $\alpha_* < \infty$, then

$$\lim_{t \to \infty} P\{z(t)=k\} = p_k^*, \quad k \in N_0,$$

where the generating function of the probability distribution $\{p_k^*, k \in N_0\}$ is equal to

$$F(s) = \exp\left\{ \int_s^1 \frac{h_*(x)}{f_*(x)} dx \right\}.$$

Proof. Since

$$|h_*(F(u,s))| \le \alpha_* e^{au} |1-s|,$$

the integral

$$\int_0^\infty h_*(F(u,s)) du \qquad (1.29)$$

converges uniformly with respect to $|s| \le 1$. Therefore there is

$$F(s) = \lim_{t \to \infty} H(t,s) = \exp\left\{ \int_0^\infty h_*(F(u,s)) du \right\}.$$

We show that

$$\int_0^\infty h_*(F(u,s)) du = \int_s^1 \frac{h_*(x)}{f_*(x)} dx. \qquad (1.30)$$

If we differentiate integral (1.29) with respect to s, we obtain:

$$\frac{d}{ds} \int_0^\infty h_*(F(u,s)) du = \int_0^\infty \frac{dh_*(F(u,s))}{dF(u,s)} \frac{\partial F(u,s)}{\partial s} du. \qquad (1.31)$$

It is well-known (see Sevast'yanov, 1971, p.27) that the generating function $F(s)$ satisfies the partial differential equation

§2.1. Classical Models of Branching Processes

$$\frac{\partial F(u,s)}{\partial u} = f_*(s) \frac{\partial F(u,s)}{\partial s}. \qquad (1.32)$$

Therefore we obtain that the integral on the right-hand side of (1.31) equals

$$\frac{1}{f_*(s)} \int_0^\infty \frac{dh_*(F(u,s))}{dF(u,s)} \frac{\partial F(u,s)}{\partial u} du.$$

But the last integral is equal to

$$\int_0^\infty \frac{\partial h_*(F(u,s))}{\partial u} du = h_*(F(\infty,s)) - h_*(F(0,s)) = -h_*(s).$$

Hence,

$$\frac{d}{ds} \int_0^\infty h_*(F(u,s)) du = -\frac{h_*(s)}{f_*(s)}$$

and equality (1.30) is true. Since

$$\lim_{t \to \infty} h_*(F(t,s)) = 0,$$

uniformly with respect to $|s| \le 1$, and

$$\lim_{t \to \infty} \frac{h_*(F(t,s))}{f_*(s)} = 0,$$

uniformly with respect to $|s| \le 1$, $s \ne 1$, then integration with respect to the parameter s is valid. The theorem is proved.

In conclusion we shall present some limit theorems for critical and supercritical processes (see Sevast'yanov, 1971).

Theorem 1.9. If a=0 and b, $\alpha_* \in (0, \infty)$, then the random variable $2Z(t)/bt$ as $t \to \infty$ converges in distribution to a random variable having the gamma-distribution with density function

$$\omega_*(x) = \left(\Gamma(2\alpha_*/b)\right)^{-1} x^{2\alpha_*/b - 1} e^{-x}.$$

Theorem 1.10. If a>0, and b, α_*, β_* are finite, then the random variable $\xi(t) = e^{-at} Z(t)$ as $t \to \infty$ converges to a random variable ξ in the mean square and

$$\psi(\tau) = E e^{i\tau\xi} = \exp\left\{\frac{1}{a} \int_0^\tau \frac{h_*(\varphi(x))}{x} dx\right\},$$

where $\varphi(x)$ is defined by the relation:

$$\frac{d\varphi(x)}{dx} = \frac{f_*(\varphi(x))}{dx} , \quad \varphi(0)=1.$$

§2.2. GENERAL BRANCHING PROCESSES WITH REPRODUCTION-DEPENDENT IMMIGRATION

In the first papers on branching processes with immigration the process $X(t)$ was either a homogeneous Poisson process or a partial sum of independent and identically distributed random variables. Later on, processes with immigration which was either a non-homogeneous with independent increments or stationary in the broad sense were studied. A further generalization was considering a "scheme of series" of arbitrary point processes as an immigration one. One can find a sufficiently full bibliography of such papers in the review of Vatutin and Zubkov (1985). However, the independence of processes of reproduction and immigration was assumed in these publications.

In this section we shall consider a general Krump-Mode-Jagers branching process with random characteristic of particles and with immigration which possibly depends on reproduction. We will assume that the evolution of processes generated by immigrating particles depends on the initial state of a random medium. A general theorem allowing us to obtain limit theorems for branching processes with immigration, when a limit distribution of the corresponding process without immigration is known, will be proved. Later on, we shall call such kind of theorems "transfer theorems for branching processes". Limit theorems for the critical process with an "individual" characteristic of particles and for a total progeny of the Bellman-Harris process will be proved as an application of the general theorem.

Proof of the general theorem uses Theorem 1.1.2 from §1.1 on convergence to a mixture of infinitely divisible distributions.

§2a. The Model

Consider a general branching process with immigration defined by a family of point processes $(\theta_k^t, \eta_k^t)_{k=1}^{\infty}$, where θ_k^t are random variables with values from the set $R_+=[0,\infty)$; η_k^t are random variables taking values from $N=\{1,2,\ldots\}$, $t \in R_+$. The variables θ_k^t and η_k^t mean the time and the size of a k-th jump of the process

$$X^t(u) = \sum_{k \geq 1} \chi\{\theta_k^t \leq u\} \eta_k^t, \qquad (2.1)$$

respectively. Here $\chi(A)$, as before, is the indicator function of an event A.

Suppose that the random number η_k^t particles immigrate at time θ_k^t, $k \in N$. Each of the immigrating particles generates a general branching process whose evolution depends on the initial state of a random environment.

Let us go over to describe the probability space, on which the branching process considered here is defined. If we enumerate simultaneously immigrating particles by $1,2,\ldots$, then the pair (i_1,i_2) corresponds to the i_2-th particle immigrating at time $\theta_{i_1}^t$. Later on we shall call the general branching process generated by particle (i_1,i_2) as "(i_1,i_2)-process".

Now, denote particles from an (i_1,i_2)-process by elements of the set $I = \bigcup_{k=0}^{\infty} N^k$. We denote an initial particle (ancestor) by 0, $N^0=\{0\}$, its direct offspring by elements of N, its children's children by elements of N^2 and so on. The interpretation, for instance, is that: $x=(x_1,\ldots,x_n) \in I$ is the x_1-th child of ... x_n-th child of the ancestor, who is labelled 0. For details see Jagers (1975).

Under our notations it is possible that particles corresponding to the same $x=(x_1,\ldots,x_n)$ exist in different (i_1,i_2)-processes. Therefore, besides its "name", we also show the (i_1,i_2)-process in which the particle lives.

We associate the set $\Omega_{i_1 i_2}^x$ with the σ-algebra $\mathcal{A}_{i_1 i_2}^x$ of its subsets to a particle x from (i_1,i_2)-process. Assume that sets $\Omega_{i_1 i_2}^x$ and $\mathcal{A}_{i_1 i_2}^x$ are such that reproduction point process $\xi_{i_1 i_2}^x(a)$ and random life-lengths $\lambda_{i_1 i_2}^x$ can be defined on spaces $(\Omega^X_{i_1 i_2}, \mathcal{A}^X_{i_1 i_2})$. Note that the $\xi^X_{i_1 i_2}(a)$ means a number of births up to and including age a. The basic sample space of (i_1, i_2) - process is

$$\left\{\Omega^I_{i_1 i_2}, \mathcal{A}^I_{i_1 i_2}\right\} = \prod_{x \in I} \left\{\Omega^X_{i_1 i_2}, \mathcal{A}^X_{i_1 i_2}\right\}.$$

Let the family of point processes $X^t(u), u, t \in R_+$, is defined on a measurable space $\{\Omega_T, \mathcal{A}_T\}$. If we put

$$\Omega_\Pi = \Omega_T \times \prod_{(i_1,i_2) \in N^2} \Omega^I_{i_1 i_2}, \quad \mathcal{A}_\Pi = \mathcal{A}_T \otimes \bigotimes_{(i_1,i_2) \in N^2} \mathcal{A}^I_{i_1 i_2},$$

then $\{\Omega_\Pi, \mathcal{A}_\Pi\}$ is the basic sample space of the branching process with immigration. Elements of the set Ω_Π (sample points) have the form:

$$\omega = \left\{\left\{\left\{\omega_x^{i_1 i_2}; x \in I\right\}; (i_1, i_2) \in N^2\right\}, \left\{\varphi^t(u), u, t \in R_+\right\}\right\},$$

where $\omega_x^{i_1 i_2}$ carries information on the life (life-length, childbearing process and so on) of the particle x from (i_1, i_2) - process, $\varphi^t(u)$ is a "trajectory" of the family of point processes $X^t(u)$.

If we denote by $\{\Omega_0, \mathcal{F}_0\}$ a measurable space on which the random environment is defined, then putting $\Omega = \Omega_0 \times \Omega_\Pi$, $\mathcal{A} = \mathcal{F}_0 \otimes \mathcal{A}_\Pi$, at last, we obtain probability space $\{\Omega, \mathcal{A}, P\}$, on which considering by us branching process is defined.

As Jagers (1975) and Nerman (1984), we introduce a random characteristic $\chi_x^{i_1, i_2}(a)$, $a \in R = (-\infty, \infty)$, defined on $\Omega^I_{i_1 i_2}$ of particles for each (i_1, i_2)-process. Assume that they take non-negative values, is equal to zero for $a < 0$ and the map $(a, \{\omega_y^{i_1 i_2}, y \in I\}) \curvearrowright \chi_x^{i_1 i_2}(a)$ is measurable with respect to $B \times \mathcal{A}^I_{i_1 i_2}$, B being the Borel algebra on R.

We define the branching process with random characteristics of particles by the following relation:

§2.2. General Branching Processes

$$z^{\chi}_{i_1 i_2}(t) = \sum_{x \in I} \chi_x^{i_1 i_2}(t - \sigma^{\chi}_{i_1 i_2}), \qquad (2.2)$$

where $\sigma^{\chi}_{i_1 i_2}$ is the birth time of the particle x from the (i_1, i_2)-process.

It is clear that the process $\{z^{\chi}_{i_1 i_2}(t), t \in R_+\}$ is measurable with respect to $\mathcal{A}^I_{i_1 i_2}$ for any pair (i_1, i_2). Let $\{B_{i_1 i_2}(t), t \in R_+\}$ be a family of events from $\mathcal{A}^I_{i_1 i_2}$. We define the branching process with immigration by the relation

$$W^{\chi}_t = \sum_{i_1=1}^{N(t)} \sum_{i_2=1}^{\eta^t_{i_1}} z^{\chi}_{i_1 i_2}(t - \theta^t_{i_1}) \chi(B_{i_1 i_2}(t - \theta^t_{i_1})), \qquad (2.3)$$

where $N(t)$ is the number of jumps up to time t, that is

$$N(t) = \sum_{k: \theta^t_k \leq t} 1.$$

Note that we have not spoken about the independence of (i_1, i_2)-processes and immigration processes until now. Presuppose now that the following basic assumptions are fulfilled:

(a) If the initial state of the random environment is fixed, immigrating particles reproduce independently according to identically distributed general Krump-Mode-Jagers processes; pairs $(z^{\chi}_{i_1 i_2}(t), \chi(B_{i_1 i_2}(t)))$ are independent and identically distributed for different pairs (i_1, i_2);

b) the family $\{\theta^t_k, k \in N, t \in R_+\}$ is measurable with respect to $\mathcal{F}_0$, and for any k, t, and j the variables η^t_k satisfy the condition:

$$\{\eta^t_k \leq j\} \in F_{kj}(t) = \prod_{l=1}^{k-1} \prod_{i=1}^{\eta^t_l} \mathcal{A}^I_{li} \times \prod_{i=1}^{j} \mathcal{A}^I_{ki} \times \mathcal{F}_0, \qquad (2.4)$$

where $\prod_{i=1}^{0} \mathcal{A}^I_{li} = \{\phi, \Omega\}$ and the direct product of σ-algebras, as in §1.1 we understand that:

$$\prod_{i=1}^{\eta} \mathcal{A}^I_{ki} = \left\{ A \in \mathcal{A} : A \cap \{\eta = j\} \in \prod_{i=1}^{j} \mathcal{A}^I_{ki} \right\}.$$

Further, the integer valued random variables $\{\eta^t_k, k \in N\}$ satisfying condition (2.4) are called **double stopping times** with respect to the $\{F_{kj}(t)\}$.

It follows from assumption (a) that the lives of different particles are independent and identically distributed; the life laws Q^x are defined on $\{\Omega^x_{i_1 i_2}, \mathcal{A}^x_{i_1 i_2}\}$, $x \in I$, and the probabilities Q^B of the events $B_{i_1 i_2}(u)$ are $\mathcal{F}_0$-measurable random variables.

The assumption (b) means that the immigration moments are determined by the initial state of the random environment and the number of particles then immigrating possibly depends on the number of particles which have immigrated up to this time, and on the branching processes generated by them.

If in considering here model σ-algebras $\mathcal{A}^i_{i_1 i_2}$, $(i_1, i_2) \in N^2$ and $\mathcal{F}_0$ are independent and variables η^t_k are $\mathcal{F}_0$-measurable for any $k \in N$, $t \in R_+$, then we have the branching process with reproduction-independent immigration.

§2b. The Main Theorem

Let

$$P_0(A) = P\{A | \mathcal{F}_0\}, \quad M_0[\xi] = M[\xi | \mathcal{F}_0], \quad \mu(t) = M_0[\xi_{i_1 i_2}(t)],$$

$$\hat{z}^x_{i_1 i_2}(t) = z^x_{i_1 i_2}(t) \chi(B_{i_1 i_2}(t)), \quad Y^x_{i_1 i_2}(t) = \hat{z}^x_{i_1 i_2}(t) - M_0\left[\hat{z}^x_{i_1 i_2}(t)\right].$$

Assume that there exists $0 < \varepsilon_0 < 1$ for which

$$\sup_t \sup_{0 \leq x \leq \varepsilon_0} M_0[\hat{z}^x(tx)] / M_0[\hat{z}^x(t)] \leq C_0, \qquad (2.5)$$

almost everywhere, where

$$\hat{z}^x(t) = \hat{z}^x_{11}(t), \quad z^x(t) = z^x_{11}(t), \quad B(t) = B_{11}(t),$$

and C_0 is some constant.

We will use the notations

$$a(t) = M_0[z^x(t) | B(t)], \quad Q(t) = P_0(B(t)).$$

Introduce the following conditions:

1) As $t \to \infty$

$$Q(t) X^t(ut) \xrightarrow{P} X(u), \qquad (2.6)$$

for any $u \in [0,1]$, where $X(u)$ is some $\mathcal{F}_0$-measurable, stochastically continuous for $u=1$ random process with non-decreasing trajectories, $X(0)=0$ and $X(1) < \infty$ almost everywhere;

§2.2. General Branching Processes

2) as $t\to\infty$

$$P_0\{z^\chi(t)<a(t)x \mid B(t)\} \xrightarrow{P} F(x), \qquad (2.7)$$

for any $x\in R_+$, where $F(x)$ is $\mathcal{F}_0$-measurable and continuous on the set $(0,\infty)$ distribution function;

3) as $t\to\infty$ the variable $Q(t)\xrightarrow{P} 0$ and

$$\frac{a(tx)}{a(t)} \xrightarrow{P} \pi_1(x), \quad \frac{Q(tx)}{Q(t)} \xrightarrow{P} \pi_2(x) \qquad (2.8)$$

for any $x\in(0,1]$ uniformly with respect to x from a set of the form $[\varepsilon,1]$, $\varepsilon>0$, where $\pi_i(x)$ are $\mathcal{F}_0$-measurable, continuous in $(0,1]$ functions such that for some (possibly depending on ε) numbers a_i, b_i

$$0<a_i \leq \inf_{\varepsilon\leq x\leq 1} \pi_i(x) \leq \sup_{\varepsilon\leq x\leq 1} \pi_i(x) \leq b_i < \infty$$

almost everywhere, and there are the integrals

$$\gamma = \int_0^1 \pi_1(1-u)\pi_2(1-u)dX(u),$$

$$H(x) = \int_0^1 \left[1-F\left(\frac{x}{\pi_1(1-u)}\right)\right]\pi_2(1-u)dX(u), \quad x>0. \qquad (2.9)$$

almost everywhere, $H(x)=0$ for $x\leq 0$.

Processes under the differential symbol have non-decreasing trajectories and trajectories of integrated processes are continuous. In this case they can be defined as stochastic Riemann-Stieltjes integrals (Liptser and Shiryaev, 1986).

Theorem 2.1. Let assumptions (a) and (b) be satisfied. Then, under the conditions (2.5)-(2.8), the variable $W_t^\chi/a(t) \xrightarrow{d} W$ as $t\to\infty$, where

$$Me^{itW} = M\exp\left\{-\int_0^\infty (e^{itx}-1)dH(x)\right\}. \qquad (2.10)$$

Remark. 2.1. The conditions of Theorem 2.1 can be satisfied for the extensive class of processes with random characteristics. They are fulfilled, for example, for a critical general branching process z_t (if we put the characteristics $\chi_x(a)=1_{[0,\lambda_x]}(a)$, where λ_x is the life-length of particle x, and $B(t)=\{z_t>0\}$), and for the total progeny Y_t of the process (characteristic $\chi_x(a)=1_{R_+}(a)$). Considering the concrete characteristics and choosing events B(t) in corresponding form, one can find different limit theorems from Theorem 2.1 and from limit theorems for branching processes without immigration.

Remark 2.2. Limit theorems under the condition of the form (2.7) were proved for the first time by Rahimov (1984a) and by S.Nagayev, M.Asadullin (1985) (in the case where the limit distribution is exponential and $B(t)=\{z_t>0\}$). In these papers processes with reproduction-independent immigration having no random characteristic and whose reproduction process is independent of the environment were considered. The condition of the form (2.6) was suggested by I.Badalbayev and A.Zubkov (1983) (they demand convergence of finite-dimensional distributions).

Remark 2.3. In this section we shall prove Theorem 2.1 for independent and identically distributed (i_1,i_2)-processes. By our methods one can obtain limit theorems for "scheme of series" in the case when distinct (i_1,i_2)-processes have distinct distributions.

Example 2.1. Let

$$B_{i_1 i_2}(u) = \begin{cases} \{\max_{1\leq s\leq u} z^\chi_{i_1 i_2}(s)\geq M,\ u\leq L,\} \\ B_{i_1 i_2}(L) \quad,\ u>L, \end{cases}$$

where L and M are some fixed positive numbers. In this case W^χ_t describes the following population process with "selection". We observe the family (daughter process) of each immigrating particle during the finite time interval [0,L]. If the total characteristic of the family does not achieve the level M until the time L, then this family will be excluded from the process (may be it can emigrate). For example, M can be interpreted as a level of accumulated wealth by an immigrating family in the demographic processes. If we consider an individual-based population process (animals, bacteria, cells) M may be the level of the number of descendants of an immigrant or the level of their production. It is clear that the events $B_{i_1 i_2}(u)$ can be chosen in different forms depending on the demands of practical problems. Branching processes with reproduction-independent group emigration were considered by A. Pakes (1986).

§2c. The Proof of the Main Theorem

Proof of Theorem 2.1. Let m_t, $t\in R_+$, be a non-negative integer-valued

§2.2. General Branching Processes

function such that

$$P\{\kappa_t=1\} \to 1, \quad \kappa_t = \chi\{N(t) \le m_t\} \tag{2.11}$$

as $t \to \infty$. We consider the sum

$$V_t = \sum_{i_1=1}^{m_t} \sum_{i_2=1}^{\hat{\eta}_{i_1}(t)} \hat{z}^{\chi}_{i_1 i_2}(t-\theta^t_{i_1}), \quad \hat{\eta}_{i_1}(t) = \chi(i_1 \le N(t))\eta^t_{i_1}. \tag{2.12}$$

We shall prove that Theorem 1.1.2 is applicable with $r=2$ to (2.12), under the conditions of Theorem 2.1.

The variables $\hat{z}^{\chi}_{i_1 i_2}(t-\theta^t_{i_1})/a(t)$ are measurable with respect to $\mathcal{F}^I_{i_1 i_2} = \mathcal{A}^I_{i_1 i_2} \times \mathcal{F}_0$ for all pairs $(i_1, i_2) \in \mathbb{N}^2$, $t \in \mathbb{R}_+$. Since $\{i_1 \le N(t)\} = \{\theta^t_{i_1} \le t\} \in \mathcal{F}_0$, condition (2.4), in which $\mathcal{A}^I_{i_1 i_2}$ are replaced by $\mathcal{F}_{i_1 i_2}$, is true for the family of random variables $\hat{\eta}_i(t)$.

It follows from (a) that

$$M\left[\hat{z}^{\chi}_{i_1 i_2}(t-\theta^t_{i_1})\Big|F^*_{i_1 i_2-1}(t)\right] = M\left[\hat{z}^{\chi}_{i_1 i_2}(t-\theta^t_{i_1})\Big|\mathcal{F}_0\right], \tag{2.13}$$

where $F^*_{i_1 i_2}(t)$ is obtained from (2.4) by substituting $\mathcal{F}_{i_1 i_2}$ for $\mathcal{A}^I_{i_1 i_2}$. Therefore, since $\eta^t_i = \Delta X^t(\theta^t_i) = X^t(\theta^t_i) - X^t(\theta^t_i-)$, we get

$$\kappa_t \sum_{i_1=1}^{m_t} \sum_{i_2=1}^{\hat{\eta}_{i_1}(t)} M\left[\hat{z}^{\chi}_{i_1 i_2}(t-\theta^t_{i_1})\Big|F^*_{i_1 i_2-1}(t)\right]$$

$$= \kappa_t \int_0^t M_0\left[\hat{z}^{\chi}(t-u)\right]dX^t(u). \tag{2.14}$$

First, we show that

$$I(t) \stackrel{\text{def}}{=} \frac{1}{a(t)} \int_0^t M_0\left[\hat{z}^{\chi}(t-u)\right]dX^t(u) \xrightarrow{P} \gamma, \tag{2.15}$$

as $t \to \infty$, where γ is defined by relation (2.9). Let $\varepsilon \in (0, \varepsilon_0)$. We represent $I(t)$ in the form

$$I(t) = \frac{1}{a(t)} \int_0^{(1-\varepsilon)t} M_0\left[\hat{z}^{\chi}(t-u)\right]dX^t(u)$$

$$+ \frac{1}{a(t)} \int_{(1-\varepsilon)t}^{t} M_o\left[\hat{z}^{\chi}(t-u)\right] dX^t(u) = I_1(t) + I_2(t). \qquad (2.16)$$

Let us consider the first term. If we use the equality $M_o\left[\hat{z}^{\chi}(t)\right] = a(t)Q(t)$, we obtain:

$$I_1(t) = \int_0^{1-\varepsilon} \pi_1(1-u)\pi_2(1-u)Q(t)dX^t(ut) + \varepsilon_1(t), \qquad (2.17)$$

where

$$\varepsilon_1(t) = \int_0^{1-\varepsilon} \left[\frac{a(t(1-u))Q(t(1-u))}{a(t)Q(t)} - \pi_1(1-u)\pi_2(1-u)\right] Q(t)dX^t(ut).$$

We conclude from conditions (2.6) and (2.8) that $\varepsilon_1(t)$ converges to zero in probability as $t \to \infty$. On the strength of (2.6) it follows from (2.17) that

$$I_1(t) \xrightarrow{P} \int_0^{1-\varepsilon} \pi_1(1-u)\pi_2(1-u)dX(u). \qquad (2.18)$$

We can see from the inequality

$$I_2(t) \leq \sup_{0 \leq x \leq \varepsilon} \frac{M_o[\hat{z}^{\chi}(tx)]}{M_o[\hat{z}^{\chi}(t)]} Q(t)[X^t(t) - X^t(t(1-\varepsilon))],$$

from conditions (2.5), (2.6) and from the choice of $\varepsilon \in (0, \varepsilon_o)$, that the second term in (2.16) is less than a variable which converges to the $C_o[X(1) - X(1-\varepsilon)]$ as $t \to \infty$ in probability. Thus, if we turn ε to zero, we obtain relation (2.15) in view of the fact that $X(u)$ is stochastically continuous at the point $u=1$.

Now we show that, under the conditions of Theorem 2.1,

$$H_t(x) \overset{\text{def}}{=} \int_0^t P_o\left\{\frac{Y^{\chi}(t-u)}{a(t)} > x\right\} dX^t(u) \xrightarrow{P} H(x), \qquad (2.19)$$

as $t \to \infty$ for $x > 0$, where $Y^{\chi}(t) = \hat{z}^{\chi}(t) - M_o[\hat{z}^{\chi}(t)]$ and $H(x)$ is the same function as in the statement of the theorem. Let $\varepsilon \in (0, \varepsilon_o)$ and

$$H_t(x) = H_t^{(1)}(\varepsilon, x) + H_t^{(2)}(\varepsilon, x), \qquad (2.20)$$

where $H_t^{(1)}$ and $H_t^{(2)}$ are integrals of the form (2.19) with integration domain $[0, t(1-\varepsilon)]$ and $[t(1-\varepsilon), t]$, respectively. If we use the following inequality

$$P_o\{Y^{\chi}(t-u) > xa(t)\} \leq (xa(t))^{-1} M_o[\hat{z}^{\chi}(t-u)],$$

§2.2. General Branching Processes

we obtain for the second term in (2.20) the estimate

$$H_t^{(2)}(\varepsilon,x) \le I_2(t)x^{-1}, \qquad (2.21)$$

where $I_2(t)$ is defined by (2.16).

It follows from condition (2.7) that, as $t\to\infty$ for any $x>0$,

$$\frac{1}{Q(t)} P_o\{\hat{z}^\chi(t) > a(t)x\} \xrightarrow{P} \hat{F}(x), \quad \hat{F}(x)=1-F(x).$$

Since $M_o[z^\chi(t)]=a(t)Q(t)$ and since function $F(x)$ is continuous for $x\in(0,\infty)$, then as $t\to\infty$

$$\sup_{a\le x\le b} R_t(x) \xrightarrow{P} 0, \quad R_t(x) = \left| \frac{1}{Q(t)} P_o\left\{ \frac{Y^\chi(t)}{a(t)} > x \right\} - \hat{F}(x) \right|,$$

where $0<a\le b<\infty$ are some numbers. Choose $\delta\in(0,a_1)$, where a_1 is the same as in the condition (2.8), and denote

$$\psi_t = \chi\left(\sup_{\varepsilon\le y\le 1} \left| \frac{a(ty)}{a(t)} - \pi_1(y) \right| \le \delta \right).$$

Then on the strength of (2.8) and the boundedness of $R_t(x)\le 2$

$$\sup_{\varepsilon\le u\le 1} R_t\left(\frac{a(t)x}{a(tu)} \right) (1-\psi_t) \xrightarrow{P} 0. \qquad (2.22)$$

But on the other hand

$$\sup_{\varepsilon\le u\le 1} R_{tu}\left(\frac{a(t)x}{a(tu)} \right) \psi_t \le \sup_u \sup_x R_{tu}(x), \qquad (2.23)$$

where $u\in[0,1]$, $x\in[a_1-\delta,b_1+\delta]$.

We obtain at last from (2.22) and (2.23) that

$$\sup_{\varepsilon\le u\le 1} R_t\left(\frac{a(t)x}{a(tu)} \right) \xrightarrow{P} 0 \qquad (2.24)$$

as $t\to\infty$ and for any fixed $x>0$.

Now we can represent the first summand in (2.20) as:

$$H_t^{(1)}(\varepsilon,x) = \int_0^{t(1-\varepsilon)} \left[1-F\left(\frac{a(t)x}{a(t-u)} \right) \right] Q(t-u)dX^t(u) + \varepsilon_2(t,x), \qquad (2.25)$$

where $\varepsilon_2(t,x)$ has the estimate

$$|\varepsilon_2(t,x)| \le \int_0^{1-\varepsilon} R_{t(1-u)}\left(\frac{a(t)x}{a(t(1-u))} \right) Q(t(1-u))dX^t(ut).$$

This inequality, together with relation (2.24) and conditions (2.6) and (2.8), shows that $\varepsilon_2(t,x)$ converges to zero as $t\to\infty$ in probability.

If we use the continuity of the function $F(x)$, $x>0$, and conditions (2.6) and (2.8), we obtain that the first term in (2.25) converges as $t\to\infty$ to

$$\int_0^{1-\varepsilon} \left[1-F\left(\frac{x}{\pi_1(1-u)}\right)\right]\pi_2(1-u)dX(u),$$

in probability. We are justified in correctness of relation (2.19) by (2.20), taking into account relations (2.21), (2.25) and the estimate of $I_2(t)$ in (2.20).

We will now proceed to verify the conditions of Theorem 1.1.2. We shall prove that condition (1.1.15) is fulfilled for $r=2$. Let us consider the sum:

$$K_t(x) = \sum_{i_1=1}^{m(t)} \sum_{i_2=1}^{\hat{\eta}_{i_1}(t)} M\left[g\left(\frac{Y^x_{i_1 i_2}}{a(t)}, x\right)\right] \left|F^*_{i_1 i_2-1}(t)\right|,$$

where

$$Y^x_{i_1 i_2}(t) = \hat{z}^x_{i_1 i_2}(t) - M_o\left[\hat{z}^x_{i_1 i_2}(t)\right], \quad x\in[0,\infty).$$

The same arguments used in the proof of (2.14) show that

$$\kappa_t K_t(x) = \kappa_t \int_0^t M_o\left[g\left(\frac{Y^x(t-u)}{a(t)}, x\right)\right]dX^t(u),$$

where $Y^x(t)=Y^x_{11}(t)$, and κ_t is defined in (2.11).

We assume that $x>0$ and choose $\varepsilon\in(0,1)$. We shall divide the interval $[0,x]$ by the points $0<x_0<x_1<\ldots<x_r=x$ such that $x_j-x_{j-1}<\varepsilon$, $x_0<\varepsilon$. Then:

$$\kappa_t K_t(x) = \kappa_t [S_1(t,r)+\varepsilon_3(t,r)+\varepsilon_4(t,r)], \qquad (2.26)$$

where

$$S_1(t,r) = \sum_{j=1}^r \frac{x_{j-1}^2}{1+x_{j-1}^2} \int_0^t P_o\left\{x_{j-1} < \frac{Y^x(t-u)}{a(t)} \leq x_j\right\}dX^t(u),$$

$$\varepsilon_3(t,r) = \int_0^t M_o\left[g\left(\frac{Y^x(t-u)}{a(t)}, x_o\right)\right]dX^t(u).$$

and, in view of the choice of the points for the partition, we have:

§2.2. General Branching Processes

$$|\varepsilon_4(t,r)| \leq 2\varepsilon x \int_0^t P_0\left\{x_0 < \frac{Y^{\chi}(t-u)}{a(t)} \leq x\right\} dX^t(u) . \quad (2.27)$$

It follows from (2.19) that the variable on the right-hand side of (2.27) converges in probability to $2\varepsilon x[H(x_0)-H(x)]$ as $t \to \infty$ for any fixed partition.

Consider $\varepsilon_3(t,r)$. We use the following equality:

$$\varepsilon_3(t,r) = \int_0^t M_0\left[g\left(\frac{Y^{\chi}(t-u)}{a(t)}, x_0\right); Y^{\chi}(t-u)<0\right] dX^t(u)$$

$$+ \int_0^t M_0\left[g\left(\frac{Y^{\chi}(t-u)}{a(t)}, x_0\right); Y^{\chi}(t-u)>0\right] dX^t(u). \quad (2.28)$$

If we divide the integration domain in (2.28) into the parts $[0,t(1-\varepsilon_1))$, $[t(1-\varepsilon_1),t)$, $\varepsilon_1 \in (0,\varepsilon_0)$, we can conclude that the first summand is less than

$$\sup_{t\varepsilon \leq u \leq t} Q(u) \int_0^{t(1-\varepsilon_1)} M_0[\hat{z}^{\chi}(t-u)] \frac{a(t-u)}{a^2(t)} dX^t(u).$$

If we use the same arguments as in the proof of relation (2.15), we obtain that the integral in this expression converges in probability to

$$\int_0^{1-\varepsilon_1} \pi_1^2(1-u)\pi_2(1-u) dX(u).$$

Therefore the integral over the interval $[0,t(1-\varepsilon_1)]$ converges to zero in probability as $t \to \infty$ for any $\varepsilon_1 \in (0,\varepsilon_0)$.

Using the definition of $a(t)$ and condition (2.5), we have the integral over the second interval being less than $C_0 Q^2(t) X^t(t)$, which converges in probability to zero as $t \to \infty$, in view of condition (2.6).

The second integral in (2.28) is majorized by

$$\frac{x_0}{a(t)} \int_0^t M_0[\hat{z}^{\chi}(t-u)] dX^t(u).$$

Since $x_0 < \varepsilon$, we also obtain that the second term in (2.28) converges to zero in probability as $t \to \infty$ on the strength of (2.15).

Thus, it is possible to choose the quantity $\varepsilon \in (0,1)$ in (2.26) such that for any $\delta > 0$

$$P\{\varepsilon_3(t,r) > \delta\} \longrightarrow 0, \quad t \to \infty. \quad (2.29)$$

Let us consider the sum $S_1(t,r)$ from (2.26). We represent it in the form:

70   II. Branching Processes with Generalized Immigration

$$S_1(t,r) = -\int_0^y \frac{x^2}{1+x^2} dH(x) + \varepsilon_5(t,r) + \varepsilon_6(r),$$

where

$$\varepsilon_5(t,r) = \sum_{j=1}^r \frac{x_{j-1}^2}{1+x_{j-1}^2} \left[\int_0^t P_o\left\{ x_{j-1} < \frac{Y^x(t-u)}{a(t)} \le x_j \right\} dX^t(u) + H(x_j) - H(x_{j-1}) \right],$$

$$\varepsilon_6(r) = \int_0^t \frac{x^2}{1+x^2} dH(x) - \sum_{j=1}^r \frac{x_{j-1}^2}{1+x_{j-1}^2} [H(x_j) - H(x_{j-1})].$$

It follows from the relation (2.19) that the variable $\varepsilon_5(t,r)$ converges to zero in probability as $t\to\infty$ for any fixed partition, and variable $\varepsilon_6(r)$ converges to zero in probability as $r\to\infty$. Thus,

$$S_1(t,r) \xrightarrow{P} -\int_0^y \frac{x^2}{1+x^2} dH(x), \quad t, r\to\infty. \qquad (2.30)$$

Taking into account relations (2.11), (2.29), (2.30) and the estimate of the variable $\varepsilon_4(t,r)$ from (2.26), we have that condition (1.1.15) of Theorem 1.1.2 is satisfied for $x\in R$; in addition

$$K(x) = -\int_0^x \frac{u^2}{1+u^2} dH(u),$$

for $x>0$ and is equal to zero for $x<0$.

In the case of $x=\infty$ we use the equality

$$\kappa_t K_t(\infty) = \kappa_t K_t(L) + \kappa_t \int_0^x M_o\left[\frac{(Y^x(t-u))^2}{a^2(t)+(Y^x(t-u))^2} ; \frac{Y^x(t-u)}{a(t)} > L \right] dX^t(u),$$

which holds for any $L>0$. Since the second summand does not exceed $\kappa_t H_t(L)$, which converges in probability to $H(L)$ by relations (2.11) and (2.19), and the integral $K(\infty)$ converges almost everywhere, then condition (1.1.15) is also fulfilled in the case $x=\infty$.

Arguments, similar to those just made, show that as $t\to\infty$

$$a(t) \int_0^t M_o\left[\frac{Y^x(t-u)}{a^2(t)+(Y^x(t-u))^2}, Y^x(t-u)\ge 0 \right] dX^t(u) \xrightarrow{P} \gamma_o \equiv -\int_0^\infty \frac{x}{1+x^2} dH(x),$$

§2.2. General Branching Processes

$$a(t)\int_0^t M_o\left[\frac{Y^\chi(t-u)}{a^2(t)+(Y^\chi(t-u))^2}; Y^\chi(t-u)<0\right]dX^t(u) \xrightarrow{P} -\gamma.$$

This, together with relation (2.15), shows that condition (1.1.16) is fulfilled and $\gamma=\gamma_o$.

It remains for us to show the fulfillment of condition (1.1.17). Using the estimate

$$\left|1-M_o e^{i\tau Y^\chi(t-u)}\right| \le 2\tau M_o[\hat{z}^\chi(t-u)],$$

we obtain the two inequalities:

$$\sup_{0\le u\le t}\left|1-M_o e^{i\tau Y^\chi(t-u)/a(t)}\right| \le 2C_o\tau Q(t) + 2\tau \sup_{\varepsilon_o\le u\le 1}\frac{a(ut)}{a(t)}$$

$$\int_0^t \left|1-M_o e^{i\tau Y^\chi(t-u)/a(t)}\right| dX^t(u) \le 2\tau I(t),$$

where C_o and ε_o are quantities from condition (2.5) and where the variable $I(t)$ is defined by (2.15). We conclude from these inequalities, condition (2.8) and relations (2.11) and (2.15) that

$$\kappa_t \sum_{i_1=1}^{m_t} \sum_{i_2=1}^{\hat{\eta}_{i_1}(t)} \left|1-M_o e^{i\tau Y^\chi_{i_1 i_2}(t-\theta^t_{i_1})/a(t)}\right|^2$$

$$= \kappa_t \int_0^t \left|1-M_o e^{i\tau Y^\chi(t-u)/a(t)}\right|^2 dX^t(u)$$

converges to zero in probability as $t\to\infty$. Hence, condition (1.1.17) is also satisfied. If we put

$$\xi_{i_1 i_2}(t) = Y^\chi_{i_1 i_2}(t-\theta^t_{i_1})/a(t)$$

in Theorem 1.1.2, we obtain the statement of Theorem 2.1. Theorem 2.1 is thus proved.

§2d. Applications of the Main Theorem

Now we shall demonstrate some applications of Theorem 2.1. First, we consider a general branching process with an **individual** characteristic. It means that the characteristic of each particle x depends only on the life of the particle x, i.e.,

$$\chi_x^{i_1 i_2}(a,\{\omega_y^{i_1 i_2}, y\in I\}) = \chi^{i_1 i_2}(a, \omega_x^{i_1 i_2}), \quad x\in I.$$

We assume that the triples $\{\lambda_{i_1 i_2}^x, \xi_{i_1 i_2}^x, \chi_x^{i_1 i_2}(a)\}$ are independent and identically distributed for different $(i_1, i_2)\in N^2$, $x\in I$; the function $\mu(t) = M\xi_{i_1 i_2}^x(t)$ is non-lattice, i.e. μ may not be supported by any lattice $\{0, d, 2d, \ldots\}$, $d>0$, $\mu(\infty)>\mu(0)$, $\mu(0)<1$. Let the characteristics $\chi_x^{i_1 i_2}(t)$ be continuous almost everywhere, and let $M[\chi_x(t)]$ be directly Riemann integrable (here and later on $\chi_x(t) = \chi_x^{11}(t)$), let $M[(\chi_x(t))^2]$ be bounded and tend to zero as $t\to\infty$.

Note that the directly Riemann integrability of the function $f(x)$ means that $\Sigma h t_n(h)$ and $\Sigma h T_n(h)$ converge absolutely for any sufficiently small $h>0$, where $T_n(h) = \sup_{nh\leq x \leq (n+1)h} f(x)$, and $t_n(h)$ is the corresponding infimum, and

$$h(\Sigma T_n(h) - \Sigma t_n(h)) \to 0, \quad h\to 0.$$

The function $f(x)$ is directly Riemann integrable, for example, if $f(x)\geq 0$ and is non-increasing and Riemann integrable in the ordinary sense.

We define by W_t^χ the total characteristic of the process with immigration and set

$$L(t) = P\{\lambda_{i_1 i_2}^x \leq t\}, \quad \sigma^2 = D\xi_{i_1 i_2}^x(\infty), \quad \bar{\chi} = \int_0^\infty M[\chi_x(u)]du, \quad \beta = \int_0^\infty u d\mu(u).$$

Assume that the reproduction of immigrating particles is independent of states of the random environment. Then the variables L, σ^2, $\bar{\chi}$, β are deterministic and one can apply well-known limit theorems to the corresponding branching process without immigration. As for the immigration process, we assume that conditions (b) of Theorem 2.1 are satisfied.

Theorem 2.2. If, under the assumptions described above, the following conditions are satisfied:

1) $a = \mu(\infty) = 1$, $0 < \sigma^2 < \infty$;

2) $P\{\chi_x(t) = 0 \mid \lambda_{i_1 i_2}^x \leq t\} = 1$;

3) $t^2(1-\mu(t)) \to 0$, $t^2(1-L(t)) \to 0$, $t\to\infty$;

§2.2. General Branching Processes

4) condition (2.6) holds with $Q(t)=2\beta/\sigma^2 t$.

Then the variable $2\beta W_t^\chi/\sigma^2\bar{\chi}t \xrightarrow{d} \omega$, where

$$Me^{it\omega} = M\exp\left\{\int_0^1 (e^{itx}-1)dG(x)\right\},$$

$$G(x) = \begin{cases} \int_0^1 e^{-x/(1-u)}(1-u)^{-1}dX(u), & x>0, \\ 0, & x\le 0. \end{cases}$$

Proof. Let us put $B_{i_1 i_2}(t) = \{z^\chi_{i_1 i_2}(t)>0\}$, $Q(t)=P\{B_{11}(t)\}$. In this case

$$\frac{a(tx)}{a(t)} = \frac{Mz^\chi(tx)}{Mz^\chi(t)} \cdot \frac{Q(t)}{Q(tx)}. \qquad (2.31)$$

It was proved in Jagers (1975, p. 51) that under the conditions 1), 3), 4) $tP\{T>t\}\to 2\beta/\sigma^2$, $t\to\infty$, where $T=\inf\{t: z^{(a)}_{[0,\lambda)}(t) = 0\}$, that is, T is the degeneration time of the process $z(t)$. Therefore, we obtain by condition 2) that

$$tQ(t) \to 2\beta/\sigma^2, \quad t\to\infty. \qquad (2.32)$$

Since

$$Mz^\chi(t) \to \bar{\chi}/\beta, \quad t\to\infty, \qquad (2.33)$$

(see Green, 1977), we have from (2.31) the fulfillment of condition (2.8) of Theorem 2.1; moreover $\pi_1(x)=x$, $\pi_2(x)=x^{-1}$.

It follows from (2.33) that condition (2.5) is fulfilled also. As for the condition (2.7), it is satisfied for the function $F(x)=1-e^{-x}$ by virtue of Green's Theorem 2 (see Green, 1977).

The statement of the theorem follows from Theorem 2.1 and the fact that due to the choice of the event $B_{i_1 i_2}(t)$

$$z^\chi_{i_1 i_2}(t)\chi(B_{i_1 i_2}(t)) = z^\chi_{i_1 i_2}(t).$$

Corollary 2.1. If conditions 1)-4) are fulfilled with $X(t)=\lambda t$, where λ is some $\mathcal{F}_0$ - measurable random variable, then the limit distribution in Theorem 2.2 is a mixture of gamma-distributions with the characteristic function

II. Branching Processes with Generalized Immigration

$$\varphi(\tau) = M \exp\left\{\lambda \int_0^\infty \frac{e^{i\tau x}-1}{x} e^{-x} dx\right\}.$$

In particular, if the λ is deterministic, then the limit distribution is gamma.

Example 2.2. We shall now consider one special version of the random characteristic χ. We put

$$\chi_x^s(u) = 1_{[0,\lambda_x \hat{} s]}(u),$$

where $\lambda_x \hat{} s = \min(\lambda_x, s)$. In this case W_t^χ is the number of particles whose age is less than s at time t. Theorem 2.2 (and Corollary 2.1) is true in this case for this characteristic and $\bar{\chi} = \int_0^s (1-L(u))du$. Such kinds of branching processes without immigration were considered by Jagers (1975).

Now let $\chi_x(u)=1_{R_+}(u)$. In this case $Y_t = Z_{11}^{1_{R_+}(a)}(t)$ is the total number of particles born in the (1,1) - process during the time t. Assume that the reproduction process $\xi_{11}^x(t)$ has only one jump at the time λ_{11}^x and that the random variables $\xi_{11}^x(\infty)$ and λ_{11}^x are independent. In this case we have a Bellman-Harris process for which $\mu(t)=aL(t)$.

If we put

$$B_{i_1 i_2}(t) = \left\{Z_{i_1 i_2}^{1_{[0,\lambda_x)}(a)}(t) > 0\right\},$$

then W_t^χ is the total number of particles which have appeared before the time t and whose descendants exist at time t.

Let

$$P\left\{\xi_{i_1 i_2}^x(\infty) \leq 1\right\} < 1, \quad L(0+)=0. \tag{2.34}$$

Theorem 2.3. If, under the assumptions above, $a=\mu(\infty)=1$, σ^2, β are finite, $t^2(1-L(t)) \to 0$, $t \to \infty$, and condition (2.6) is fulfilled with $Q(t)=2\beta/\sigma^2 t$, then the variable

$$\frac{3\beta^2}{\sigma^2 t^2} W_t^\chi \xrightarrow{d} W$$

as $t \to \infty$, where

§2.2. General Branching Processes

$$Me^{itW} = M\exp\left\{-\int_0^\infty (e^{itx}-1)dR(x)\right\},$$

$$R(x) = \begin{cases} \int_0^1 \left[1-F\left(\frac{x}{(1-u)^2}\right)\right](1-u)^{-1}dX(u), & x>0, \\ 0, & x\leq 0, \end{cases}$$

and $F(y)$ is a distribution function with density

$$f(y) = \frac{\partial^2}{\partial x^2}\theta_4\left(\frac{x}{\sqrt{24\beta}}, \frac{y}{\sqrt{6\beta}}\right)\bigg|_{x=0}, \tag{2.35}$$

$$\theta_4(x,y) = \frac{1}{\sqrt{\pi y}}\sum_{i=-\infty}^{\infty} \exp\left\{-\frac{(x+i/2)^2}{y}\right\}, \quad y>0.$$

Remark 2.4. The distribution with density $f(y)$ can be defined by the Laplace transform (see Pakes, 1971a, 1972)

$$\int_0^\infty e^{-\lambda x}dF(x) = \sqrt{6\beta\lambda}\,\operatorname{cosec}(h[\sqrt{6\beta\lambda}]). \tag{2.36}$$

Proof of Theorem 2.3. We shall prove that the condition (2.8) is satisfied. We have from Theorem 1 of Weiner (1972)

$$t^{-2}a(t) \to \frac{\sigma^2}{3\beta^2}, \quad t^{-1}M\left[z_{i_1 i_2}^{1_{R_+}(a)}(t); B_{i_1 i_2}(t)\right] \longrightarrow \frac{2}{3\beta}. \tag{2.37}$$

It follows from (2.32) and (2.37) that the conditions (2.5) and (2.8) of Theorem 2.1 are fulfilled with $\pi_1(x)=x^2$, $\pi_2(x)=x^{-1}$.

Further we use the following theorem from Pakes's paper (1972).

Theorem A. If for a Bellman-Harris process conditions (2.34) are fulfilled, $a=1$, the quantities σ^2, β are finite, and $t^2(1-L(t))\to 0$, $t\to\infty$, then

$$\lim_{t\to\infty} P\left\{Y_t < t^2 x \mid z_t > 0\right\} = F_*(x) \equiv F\left(\frac{3\beta^2}{\sigma^2}x\right),$$

where $F(x)$ is the distribution with density (2.35).

Theorem A shows that condition (2.7) of Theorem 2.1 is fulfilled for limit function $F(x)$. Theorem 2.3 is proved.

Corollary 2.2. If under the conditions of Theorem 2.3 the condition (2.6) holds for the function $Q(t)\sim 2\beta/\sigma^2 t$, $t\to\infty$, and the process $X(t)=\lambda t$, where λ is some $\mathscr{F}_0$ - measurable random variable, then the limit distribution in Theorem

2.3 has the following characteristic function

$$\psi(\tau) = M \exp\left\{\frac{\lambda}{2} \int_0^\infty \frac{e^{i\tau x}-1}{x}(1-F(x))dx\right\},$$

where $F(x)$ the same distribution, as in Theorem 2.3.

§2.3. DISCRETE-TIME PROCESSES

In §2.2 the general branching process in which the lives of immigrating particles depend on the initial state of random environment, and is independent of the state of the environment in other instances of time, was studied. Now we will investigate discrete-time branching processes with immigration depending on reproduction in which the evolution of processes generated by immigrating particles depends on the state of the random environment at the immigration time. We also consider the case when the number of immigrating particles, optionally, is not measurable with respect to the σ-algebra generated by reproduction processes.

§3a. The Model

Let $\{\mu_{ki}(n), k, i \in N\}$ be (not necessarily independent) discrete-time branching processes and let $f_{ki}(s)$, $|s| \leq 1$, be the offspring generating functions of of these processes. We assume that the generating functions $f_{ki}(s)$, $i \geq 1$, are $\mathcal{F}_k^*$- measurable random variables for each $k \in N_0$ and fixed s, where $\mathcal{F}_k^*$ are some σ-algebras such that the state of random environment in time k is measurable with respect to $\mathcal{F}_k^*$. We put $\mathcal{F}_0 = \prod_{k=0}^\infty \mathcal{F}_k^*$, we denote by $\eta(k,n)$ the number of immigrating particles at time k and by $Z(n)$ the process with immigration, $Z(0)=0$.

It is easy to see that the process $Z(n)$ can be represented in the form

$$Z(n) = \sum_{k=1}^{n} \sum_{i=1}^{\eta(k,n)} \mu_{ki}(n-k), \tag{3.1}$$

where $\mu_{ki}(n)$ is the process generated by the i th particle immigrating at time k.

§2.3. Discrete-Time Processes

Thus, a branching process corresponds to each pair $(k,i) \in N^2$. As in §2.2, we denote the particles from (k,i)-processes by finite ordered tuples of positive integers: the tuple $(0) = N^0$ is associated with the initial particle (immigrant), and the tuple $x' = (x_1, \ldots, x_k, j)$ is associated with the jth direct offspring of the particle $x = (x_1, \ldots, x_k)$. Then the elements of the set

$$I_n = \bigcup_{i=0}^{n} N^i$$

correspond to the particles in generations $1, 2, \ldots, n$.

Let ω_x^{ki} be the offspring number of a particle x from the (k,i)-process and let

$$\mathcal{F}_{ki}(n) = \sigma(\omega_x^{ki}, x \in I_{n-k})$$

be the σ-algebra generated by the evolution of (k,i)-process up to time t. We assume that if the state of the random environment at the immigration time is fixed, immigrating particles reproduce independently according to the Galton-Watson process. For any j let the numbers of immigrating particles $\eta(k,n)$ satisfy the condition

$$\{\eta(k,n) \leq j\} \in F_{kj}(n) = \prod_{l=1}^{k-1} \prod_{i=1}^{\eta(l,n)} \mathcal{F}_{li}(n) \times \prod_{i=1}^{j} \mathcal{F}_{ki}(n) \times \mathcal{F}_0, \qquad (3.2)$$

i.e., it is a double stopping time with respect to the $F_{kj}(n)$ (see §2.2).

If $\{\eta(k,n), k, n \in N\}$ is the double stopping time, the number of immigrating particles at time k can arbitrarily depend on the evolution of processes generated by particles immigrating up to time k. It is clear that this dependence can be different in special cases. Now we consider two examples of such processes.

Example 3.1. Let $\{\xi_k, k \in N_0\}$ be independent random variables taking non-negative integers. We define the immigration by the following relations: $\eta(0) = \xi_0$,

$$\eta(k) = \begin{cases} 0, & Z^*(k) \notin B, \\ \xi_k, & Z^*(k) \in B, \end{cases}, \quad Z^*(k) = \sum_{l=0}^{k-1} \sum_{j=1}^{\eta(l)} \mu_{lj}(k-1), \quad k \geq 1,$$

where $B \subseteq N_0$. In particular if $B = \{0\}$, then $Z(n)$ is a branching process with immigration, taking place when $Z(n) = 0$. Such a model of branching processes was considered by Foster (1971), Pakes (1971b), Sato (1975), Yamazato (1975), and Mitov and Yanev (1984).

Example 3.2. Let now the immigration process be defined by the relation

$$\eta(k,n) = \max \{i: \sum_{j=1}^{i} \mu_{kj}(n-k) \leq \xi_k x_n\}, \tag{3.3}$$

where $x_n > 1$, $n \in N$, are positive numbers, ξ_k, $k \in N_o$, are (dependent in general) $\mathcal{F}_o$-measurable random variables. In this case $Z(n)$ is a process in which approximately the same number of offspring of each "group" of immigrants are alive at any time. Below we will prove a limit theorem for such processes.

We put, as before,

$$M_{kjn}[\xi] = M[\xi | F_{kj}(n)].$$

Under our assumptions for any $(k,j,n) \in N^2$

$$M_{kj-1n} e^{it\mu_{kj}(n-k)} = M\left[e^{it\mu_{kj}(n-k)} | \mathcal{F}_o\right] = f_{kj}^{(n-k)}(e^{it}), \tag{3.4}$$

where $f_{kj}^{(n)}$ is the nth iteration of $f_{kj}(s)$ and processes $\mu_{kj}(\cdot)$ are conditionall independent, provided the states of random environment are fixed.

Using (3.4) by simple calculation we find that

$$M[\mu_{ki}(n) | \mathcal{F}_o] = a_{ki}^n, \qquad a_{ki} = M[\mu_{ki}(1) | \mathcal{F}_o],$$

$$D[\mu_{ki}(n) | \mathcal{F}_o] =$$

$$= D[\mu_{ki}(1)|\mathcal{F}_o] \left[\frac{a_{ki}^n(1-a_{ki}^n)}{a_{ki}(1-a_{ki})} \chi(a_{ki} \neq 1) + n\chi(a_{ki}=1) \right]. \tag{3.5}$$

For simplicity we shall assume that $f_{ki}(s)$ coincide for different i. We introduce the notations $a_{ki} = a_k$, $b(k) = D[\mu_{ki}(1)|\mathcal{F}_o]$, define the events

$$\mathcal{B}_1(k) = \{a_k < 1\}, \quad \mathcal{B}_2(k) = \{a_k = 1\}, \quad \mathcal{B}_3(k) = \{a_k > 1\},$$

and assume that

$$\sup_k a_k < \infty, \quad 0 < \inf_k b(k) \leq \sup_k b(k) < \infty. \tag{3.6}$$

§3b. Limit Theorems for Discrete-Time Processes

Let the following conditions hold as $n \to \infty$ for some sequence of positive numbers A_n and C_n:

§2.3. Discrete-Time Processes

$$\frac{1}{C_n} \sum_{k=1}^{n} \eta(k,n)\chi(\mathcal{B}_2(k)) - A_n \xrightarrow{P} \gamma, \qquad (3.7)$$

$$\frac{1}{C_n^2} \sum_{k=1}^{n} \eta(k,n)\chi(\mathcal{B}_2(k))b(k)(n-k) \xrightarrow{P} \sigma^2, \qquad (3.8)$$

$$\frac{1}{C_n} \sum_{k=1}^{n} \chi(\mathcal{B}_1(k)\cup\mathcal{B}_3(k))a_k^{n-k}\eta(k,n) \xrightarrow{P} 0, \qquad (3.9)$$

where σ^2, γ are $\mathcal{F}_o$ - measurable random variables and $\sigma^2 > 0$ almost everywhere.

Theorem 3.1. If $\eta(k,n)$ are double stopping times with respect to $\{F_{kj}(n)\}$, conditions (3.6)-(3.9) are satisfied and

$$\sum_{k=1}^{n} \eta(k,n)\chi(\mathcal{B}_2(k))M\left[\hat{g}\left(\frac{\mu_{k1}(n-k)-1}{C_n}\right)\varepsilon\middle|\mathcal{F}_o\right] \xrightarrow{P} 0, \qquad (3.10)$$

as $n\to\infty$ for any $\varepsilon>0$, then

$$Z(n)/C_n - A_n \xrightarrow{d} Z,$$

where $Me^{itZ} = Me^{it\gamma - t^2\sigma^2/2}$.

Remark 3.1. In the case $f_k(s) \equiv f(s)$ and $\mathcal{F}_o = \{\phi, \Omega\}$ condition (3.10) is satisfied, for example, if $M\mu^3(1) < \infty$ and $n^{-1}C_n \to \infty$, $n\to\infty$. Hence Theorem 3.1 relates to the case of increasing immigration. Theorem 3.2 proved below includes the case of an immigration process close to the stationary process. Through these arguments analogous theorems for processes with decreasing immigration can be obtained. In this connection it is necessary to use Lemma 1.3.1 instead of the theorems from §1.1.

Remark 3.2. It follows from conditions (3.7)-(3.9) that the asymptotic behavior of the process is defined by the particles generating critical processes. However, by corresponding changes in these conditions, analogous theorems for subcritical and supercritical processes can also be obtained.

Proof of Theorem 3.1. We consider the relation

$$\frac{Z(n)}{C_n} = W_1(n) + W_2(n) + W_3(n), \qquad (3.11)$$

where

$$W_j(n) = \frac{1}{C_n} \sum_{k=1}^{n} \chi(\mathcal{B}_j(k)) \sum_{i=1}^{\eta(k,n)} \mu_{ki}(n-k), \quad j=1,2,3.$$

We prove that the conditions of Theorem 1.1.3 are fulfilled for $W_2(n)$

with r=2. Let T be some $\mathcal{F}_o$ - measurable random variable such that $T \geq \sigma^2+1$ almost everywhere and

$$T_n = \frac{1}{C_n^2} \sum_{k=1}^{n} \eta(k,n)\chi(\mathcal{B}_2(k))b(k)(n-k). \quad (3.12)$$

Consider the inequality

$$P\{T_n > T\} \leq P\{T_n > \sigma^2+1, \ |T_n - T| < \varepsilon\} + P\{|T_n - T| \geq \varepsilon\}. \quad (3.13)$$

It is obvious that the first term on the right-hand side equals zero for any sufficiently small ε. The second term converges to zero as $n \to \infty$ for any $\varepsilon > 0$ by condition (3.8). Hence the condition (1.1.27) of Theorem 1.1.3 is satisfied.

Since for any $\varepsilon > 0$

$$\frac{1}{C_n^2} \max_{1 \leq k \leq n} \max_{1 \leq i \leq \eta(k,n)} D(\mu_{ki}(n-k)\chi(\mathcal{B}_2(k))|\mathcal{F}_o)$$

$$\leq \varepsilon^2 + \sum_{k=1}^{n} \eta(k,n)\chi(\mathcal{B}_2(k))M\left[\hat{g}\left(\frac{\mu_{k1}(n-k)-1}{C_n}, \varepsilon\right)\bigg|\mathcal{F}_o\right], \quad (3.14)$$

we obtain the fulfillment of condition (1.1.28) from (3.10).

Finally we prove the fulfillment of condition (1.1.30). Using (3.8) and (3.10) we find that

$$\sum_{k=1}^{n} \eta(k,n)\chi(\mathcal{B}_2(k))M\left[g\left(\frac{\mu_{k1}(n-k)-1}{C_n}, \varepsilon\right)\bigg|\mathcal{F}_o\right] \xrightarrow{P} \sigma^2 \quad (3.15)$$

for any $\varepsilon > 0$, where $g(\xi,x) = \xi^2 \chi\{\xi < x\}$. Hence condition (1.1.30) is also fulfilled; moreover $K(x)=0$, $x<0$; $K(x)=\sigma^2$, $x \geq 0$.

Thus, Theorem 1.1.3 is applicable to $W_2(n)$. The other terms in (3.11) converge to zero in probability as $n \to \infty$ by condition (3.9) and Chebyshev's inequality. Theorem 3.1 is proved.

In order to state the next theorem for $Z(n)$, we denote

$$\rho_n(x,y) = n(1-x)\chi(\mathcal{B}_2([nx])P\left\{\frac{\mu_{[nx]}([n(1-x)])}{n(1-x)} > y \bigg| \mathcal{F}_o\right\},$$

$$T_n(x) = \frac{1}{n} \sum_{i=1}^{[nx]} \eta(i,n)\chi(\mathcal{B}_2(i)),$$

and introduce new conditions. Let, as $n \to \infty$,

$$\rho_n(x,y) \xrightarrow{P} \frac{1}{r(x)} e^{-y/r(x)} \quad (3.16)$$

§2.3. Discrete-Time Processes

uniformly with respect to $x \in (0,a]$ for any a and y; the random variable $r(x)$ is $\mathcal{F}_o$ - measurable for any x;

$$T_n(x) \xrightarrow{P} T(x) \qquad (3.17)$$

for $0 \leq x \leq 1$, where $T(x)$ is some $\mathcal{F}_o$-measurable, stochastically continuous for $x=1$ random process with non-decreasing trajectories, $T(0)=0$ and $T(1)<\infty$ almost everywhere.

We introduce the random function

$$F(x) = \begin{cases} \int_0^1 \exp\left\{-\frac{x}{r(u)(1-u)}\right\} \frac{dT(u)}{r(u)(1-u)}, & x>0, \\ 0, & x \leq 0. \end{cases}$$

Theorem 3.2. Let $\eta(k,n)$ be double stopping times with respect to $\{F_{kj}(n)\}$, and let conditions (3.6) and (3.9) be satisfied for $C_n = n$. Then, under conditions (3.16) and (3.17),

$$Z(n)/n \xrightarrow{d} Z$$

as $n \to \infty$, where

$$Me^{itZ} = M \exp\left\{\int_0^\infty (e^{itx}-1)dF(x)\right\}.$$

Proof. Consider the relation (3.11) with $C_n = n$. By remarks from §1.1 it follows that, if we set

$$X_{kN_n+j}(n) = \mu_{k+1\,j}(n-k), \quad \nu_{kN_n+j}(n) = \chi(j \leq \eta_{k+1}(n)), \quad j=1,\ldots,N_n,$$

$k=0,1,\ldots,n-1$, then

$$\chi(\overline{M}_n)W_2(n) = \chi(\overline{M}_n)W_2^*(n), \qquad (3.18)$$

where N_n is a sequence of integers such that

$$P\{M_n\} \to 0, \; n \to \infty, \; M_n = \{\max_{1 \leq k \leq n} \eta(k,n) \geq N_n\}, \; \overline{M}_n = \Omega \setminus M_n,$$

and

$$W_2^*(n) = \frac{1}{n} \sum_{k=1}^n \chi(\mathcal{B}_2(k)) \sum_{j=1}^{N_n} \nu_{kN_n+j}(n) X_{kN_n+j}(n-k).$$

We show that the random variables $W_2^*(n)$ satisfy the conditions of Statement 1.1.3. Indeed, since $b(k)$ are uniformly bounded, it follows from (3.17) that conditions (1.1.31)-(1.1.33) hold, in addition $\gamma = T(1)$ and for T

II. Branching Processes with Generalized Immigration

one can take an arbitrary random variable such that $T \geq T(1)+1$.

It remains to prove the fulfillment of condition (1.1.34). Consider the sum

$$S_1(n) = \sum_{k=1}^{n} \hat{\eta}_k(n) \chi(\mathcal{B}_2(k)) M\left[(\hat{\mu}_k(n-k)/n)^2 \chi(\hat{\mu}_k(n-k)/n < y) \big| \mathcal{F}_0\right], \quad (3.19)$$

where $\hat{\eta}_k(n) = \eta(k,n)\chi(\mathcal{B}_2(k))$, $\hat{\mu}_k(n-k) = \mu_{k1}(n-k)-1$.

Assume that $y>0$ and choose $\varepsilon \in (0,1)$. We divide the interval $[0,y]$ into parts by the points $0 < x_0 < \ldots < x_m = y$ such that $x_j - x_{j-1} < \varepsilon$, $x_0 \leq \varepsilon$. Then

$$S_1(n) = S_2(n,m) + \varepsilon_1(n,m) + \varepsilon_2(n,m), \quad (3.20)$$

where

$$S_2(n,m) = \sum_{j=1}^{m} x_{j-1}^2 \sum_{k=1}^{n} \hat{\eta}_k(n) P\left\{x_{j-1} < \hat{\mu}_k(n-k)/n \leq x_j \big| \mathcal{F}_0\right\},$$

$$|\varepsilon_1(n,m)| \leq 2\varepsilon y \sum_{j=1}^{m} \sum_{k=1}^{n} \hat{\eta}_k(n) P\left\{x_{j-1} < \hat{\mu}_k(n-k)/n \leq x_j \big| \mathcal{F}_0\right\},$$

$$\varepsilon_3(n,m) = \sum_{k=1}^{n} \hat{\eta}_k(n) M\left[(\hat{\mu}_k(n-k)/n)^2 \chi(-1/n \leq \hat{\mu}_k(n-k)/n \leq x_0) \big| \mathcal{F}_0\right].$$

We show that

$$S_3(n) = \sum_{k=1}^{n} \hat{\eta}_k(n) P\left\{\hat{\mu}_k(n-k) > xn \big| \mathcal{F}_0\right\} \xrightarrow{P} F(x) \quad (3.21)$$

as $n \to \infty$ for any $x>0$. To this end we represent $S_3(n)$ in the form:

$$S_3(n) = S_3'(n) + S_3''(n) + S_3'''(n), \quad (3.22)$$

where

$$S_3'(n) = \sum_{k=1}^{[n(1-\varepsilon)]} \exp\left\{-\frac{xn}{r(k/n)(n-k)}\right\} \frac{\hat{\eta}_k(n)}{r(k/n)(n-k)},$$

$$S_3''(n) = \sum_{k=1}^{[n(1-\varepsilon)]} \left(P\left\{\hat{\mu}_k(n-k) > xn \big| \mathcal{F}_0\right\}(n-k) - \frac{1}{r(k/n)} \exp\left\{-\frac{xn}{r(k/n)(n-k)}\right\}\right) \frac{\hat{\eta}_k(n)}{(n-k)},$$

$$S_3'''(n) = \sum_{k=[n(1-\varepsilon)]+1}^{n} \hat{\eta}_k(n) P\left\{\hat{\mu}_k(n-k) > xn \big| \mathcal{F}_0\right\}.$$

We estimate the terms of (3.22) in reverse order. Using Chebyshev's

§2.3. Discrete-Time Processes

inequality we obtain that, as $n \to \infty$,

$$S_3'''(n) \leq \frac{1}{x}(T_n(1)-T_n(1-\varepsilon)) \xrightarrow{P} \frac{1}{x}(T(1)-T(1-\varepsilon)). \qquad (3.23)$$

Representing the second term in the form

$$S_3''(n) = \sum_{k=1}^{[n(1-\varepsilon)]} \left(P_{[n(1-k/n)]}(k/n, x/(1-k/n)) - \frac{1}{r(k/n)} \exp\left\{-\frac{x}{r(k/n)(1-k/n)}\right\} \right) \frac{\hat{\eta}_k(n)}{(n-k)},$$

we can see that the estimate

$$|S_3''(n)| \leq \sup_m \sup_t \sup_z \left| \rho_m(t,z) - \frac{1}{r(t)} \exp\{z/r(t)\} \right| \varepsilon^{-1} T_n(1-\varepsilon)$$

is true, where $m \in [n\varepsilon, n]$, $t \in [0, 1-\varepsilon]$, $z \in [x, x\varepsilon^{-1}]$. Hence from condition (3.16) it follows that

$$S_3''(n) \xrightarrow{P} 0 \qquad (3.24)$$

as $n \to \infty$ for any x and ε. Finally, since $T_n(x) \xrightarrow{P} T(x)$ as $n \to \infty$, the sum

$$S_3'(n) = \sum_{k=1}^{[n(1-\varepsilon)]} \exp\left\{-\frac{xn}{r(k/n)(1-k/n)}\right\} \frac{T_n(k/n)-T_n((k-1)/n)}{r(k/n)(1-k/n)}$$

converges in probability, as $n \to \infty$, to the integral

$$\int_0^{1-\varepsilon} \exp\left\{-\frac{x}{r(u)(1-u)}\right\} \frac{dT(u)}{r(u)(1-u)}$$

for any $\varepsilon \in (0,1)$ and any fixed x. Hence from (3.22)-(3.24) we obtain (3.21) in view of the stochastic continuity of $T(x)$ at the point $x=1$.

It follows from (3.21) that the quantity which estimates $|\varepsilon_1(n,m)|$ converges in probability to $2\varepsilon(F(x_0)-F(y))$ for a fixed partition as $n \to \infty$. Then, using the equality

$$\{-1/n \leq \hat{\mu}_k(n-k)/n \leq x_0\} = \{\hat{\mu}_k(n-k) = -1\} \cup \{0 \leq \hat{\mu}_k(n-k)/n \leq x_0\},$$

we obtain that $\varepsilon_2(n,m)$ is not greater than the quantity convergent to $\varepsilon T(1)$ in probability.

It is possible to represent $S_1(n)$ from (3.20) in the form

$$S_1(n) = -\int_0^y x^2 dF(x) + \sum_{i=1}^{4} \varepsilon_i(n,m), \qquad (3.25)$$

where $\varepsilon_1(n,m)$ and $\varepsilon_2(n,m)$ are defined in (3.20), and

$$\varepsilon_3(n,m) = \int_0^y x^2 dF(x) - \sum_{j=1}^{m} x_{j-1}^2 [F(x_j) - F(x_{j-1})],$$

$$\varepsilon_4(n,m) = S_2(n,m) - \sum_{j=1}^{m} x_{j-1}^2 [F(x_{j-1}) - F(x_j)].$$

Using (3.25) and the estimates of quantities from (3.25) we obtain that it is possible to choose $\varepsilon > 0$ such that

$$\lim_{n \to \infty} P\left\{ \left| S_1(n) + \int_0^y x^2 dF(x) \right| > \delta \right\} = 0$$

for any $\delta > 0$, i.e., condition (1.1.34) of Statement 1.1.3 is satisfied. In addition,

$$K(y) = - \int_0^y x^2 dF(x).$$

In order to prove the assertion of the theorem it is sufficient to see that

$$\gamma = T(1), \quad \int_0^\infty x dF(x) = T(1)$$

and that the second and the third terms of (3.11) converge to zero in probability as $n \to \infty$ in view of condition (3.9). The theorem is proved.

§3c. Some Examples

In contrast to the traditional processes with immigration, this scheme allows immigration of the "subcritical" and "supercritical" particles, besides the "critical" ones. If condition (3.9) is satisfied, then these particles cannot influence the behavior of the process. It is interesting to investigate the limits of the possibility of neglecting this influence.

Example 3.3. Let $\mathcal{F}_0 = \{\phi, \Omega\}$, the variables $\eta(k,n) = \alpha(k) = \alpha(1)$, $k \geq 1$, be deterministic and let the supercritical processes have the same mean value of offspring $a > 1$. It is clear that the immigration of supercritical particles in the beginning of the process implies a violation of condition (3.9). But if they immigrate at the end, for example, in the interval $[n - L(n), n]$, then condition (3.9) of Theorem 3.2 can be fulfilled. In fact, then

§2.3. Discrete-Time Processes

$$\frac{1}{n}\sum_{k=1}^{n}\alpha(k)a^{n-k}\chi(\mathcal{B}_3(k)) \sim \frac{\alpha(1)}{n}\sum_{k=n-L(n)}^{n}a^{n-k} \sim C\frac{a^{L(n)}}{n}, \quad n\to\infty,$$

where C is a constant. Thus the supercritical particles immigrating in the interval $[n-C_o\log_a n, n]$ cannot change the asymptotic behavior of the process if $C_o<1$. But if at least one particle immigrates in the interval with $C_o\geq 1$, then the behavior of the process can be changed.

The same arguments applied to subcritical particles show that they cannot influence the asymptotic behavior of the process in the case of stationary immigration. But if a large number of such particles have immigrated (increasing immigration) before the observation moment, then the behavior of the process can change. The following example concerns a case where the critical processes have different distributions of the number of offspring.

Example 3.4. Let $\mathcal{F}_o=\{\phi,\Omega\}$, $T(u)=\lambda u$, where λ is a positive constant, and let $f_k(s)$ have the following two forms:

$$f_{2k+1}(s)=f_1(s), \quad f_{2k}(s)=f_o(s), \quad k=0,1,\ldots.$$

Then from the limit theorem for critical processes (see §2.1, Theorem 2.1.4) it follows that

$$\rho_n(2k+1,y) \to \frac{2}{b(1)}e^{-2y/b(1)}, \quad \rho_n(2k,y) \to \frac{2}{b(0)}e^{-2y/b(0)}$$

as $n\to\infty$, $k=0,1,\ldots$. Simple calculations show that in this case

$$F(x) = \lambda \sum_{n=0}^{\infty} \frac{2}{b(n)} \int_{n+1}^{n+2} e^{-2xu/b(n)} \frac{du}{u},$$

and the limit characteristic function in Theorem 3.2 has the form

$$Me^{itz}=\exp\left\{2\lambda\int_0^\infty \frac{e^{itx}-1}{x}\left[\frac{e^{-2x/b(0)}}{b(0)(1+e^{-2x/b(0)})} + \frac{e^{-4x/b(1)}}{b(1)(1+e^{-2x/b(1)})}\right]dx\right\}.$$

We now present a corollary of Theorem 3.2. Let, as before, $\mathcal{F}_o=\{\phi,\Omega\}$, let $\eta(k,n)$ be a double stopping time with respect to the system $\{\mathcal{F}_o \times \prod_{j=1}^{i} \mathcal{F}_{jk}(n)\}$ for any k, and in addition they have the same distribution for different k and n, $f_k(s)=f(s)$, $k=0,1,\ldots$.

From the assumption $\mathcal{F}_o=\{\phi,\Omega\}$ it follows that the immigrated particles reproduce independently of one another. Therefore the variable $\eta(k,n)$ and the processes $\{\mu_{ij}(n), (i,j)\in N^2, i\neq k\}$ are independent.

Corollary 3.1. If $f'(1)=1$, $b=f''(1)/2\in(0,\infty)$ and $\alpha=M\eta(k,n)\in(0,\infty)$, then $2Z(n)/bn \xrightarrow{d} Z$ as $n\to\infty$, where $Me^{itz}=(1+it)^{-\alpha/b}$.

Corollary 3.1 shows that the well-known theorem (Sevast'yanov (1971)) on convergence to the gamma-distribution is valid without additional conditions when the number of immigrating particles is a stopping time with respect to the system of σ-algebras generated by the lives of these particles and their offspring.

We now consider a corollary of Theorem 3.1. Let the processes $\mu_{ki}(n)$ be independent of the random environment. Then $f_k(s)=f(s)$, $k\in N$, are deterministic. Consider an immigration process whose jumps are defined by relation (3.3). Assume in addition that the ξ_k, $k\in N_o$, have an exponential distribution with parameter 1 (this assumption considerably simplifies calculations). Then $\eta(k,n)$ is a renewal process stopped at the random time $\tau_k(n)=\xi_k x_n$. The distribution of the stopping time is an exponential with parameter $x_n^{-1}\in(0,1)$.

It is known (see, for example, The Handbook on Prob. Theor. and Math Statist. Kiev (1978), p.135) that

$$Ms^{\eta(k,n)} = \frac{1-f^{(n-k)}(e^{-x_n^{-1}})}{1-f^{(n-k)}(e^{-x_n^{-1}})s}. \qquad (3.26)$$

We put

$$K_{l,m}(n) = cov(\eta(e,n),\eta(m,n)), \quad \hat{n}=n^2\max(n,x_n),$$

$$\hat{K}(n) = \sum_{l,m=1}^{n} (n-e)(n-m)K_{l,m}(n).$$

Corollary 3.2. If $f'(1)=1$, $b=f''(1)\in(0,\infty)$, $f'''(1)<\infty$ and $\hat{K}(n)/\hat{n}\to 0$ as $n\to\infty$, then variable $(Z(n)-\sum_{k=1}^{n}\eta(k,n))/\sqrt{B_n}$ is asymptotically normal with parameters $(0,1)$, where

$$B_n = \frac{b^2 n^3}{6} + \frac{bn^2}{2}\left(1-e^{-x_n^{-1}}\right)^{-1}.$$

The corollary follows from Theorem 3.1 by relation (3.26).

§3d. Randomly Stopped Immigration

We now consider the case where the immigration process depends on reproduction but condition (3.2) is not satisfied. Let the random variables $\eta(k,n)$ and processes $\mu_{kj}(n)$ be not necessarily independent and $\zeta(n)$ be a sequence of non-negative integer-valued random variables. We put

$$X(n,\zeta(n)) = \sum_{k=0}^{\zeta(n)} \sum_{j=1}^{\eta(k,n)} \mu_{kj}(n-k), \qquad (3.27)$$

where $\mu_{kj}(n-k) \equiv 0$, $\eta(k,n) \equiv 0$ for $k>n$.

We note that no independence of the variables $\zeta(n)$, $\eta(k,n)$ and $\mu_{kj}(n-k)$ is assumed. Therefore it is possible to interpret $\zeta(n)$ as a random "switching" time of immigration depending on branching processes generated by particles immigrating up to this time. The simple examples of such variables are

$$\zeta_1(n) = \min\left\{i: \sum_{k=0}^{i} \sum_{j=1}^{\eta(k,n)} \mu_{kj}(n-k) \geq x_n\right\},$$

$$\zeta_2(n) = \max\left\{i: \sum_{k=0}^{i} \sum_{j=1}^{\eta(k,n)} \mu_{kj}(n-k) \leq y_n\right\},$$

where x_n and y_n are positive numbers.

We will prove a limit theorem for the process (3.27). In this connection we use the results of §1.2 obtained for a sum of randomly indexed processes.

Let $m(n)$, $l(k,n)$, $k,n \in N$, be some functions taking positive integer values. Assume that the processes $\{\mu_{jk}(n)\}$ and variables $\varphi(n)$, $\eta(k,n)$ are independent of the random environment and that $\mu_{kj} \in N_o$, $j \in N$ are independent and identically distributed discrete-time branching processes.

Introduce the following conditions:

1) there exists $M \in (1,\infty)$ such that for any $\delta > 0$

$$\lim_{n \to \infty} P\left\{\max_{0 \leq k \leq Mm(n)} \left|1 - \frac{\eta(k,n)}{l(k,n)}\right| > \delta\right\} = 0;$$

2) $\zeta(n)/m(n) \to 1$, $m(n)/n \to C \in [0,1]$, $n \to \infty$;

3) for some normalizing functions $g(x) > 0$, $x \in [0,\infty)$,

II. Branching Processes with Generalized Immigration

$$\lim_{n\to\infty} \frac{1-Me^{-\tau g(n)\mu_{ij}(n)}}{g(n)} = \varphi(\tau), \quad \tau>0,$$

where $1-\varphi(\tau)$ is the Laplace transform of some distribution, $\sup_n M\mu_{ij}(n) \leq C_0$;

4) there is $\delta_0 > 0$ for which

$$\lim_{n\to\infty} g(n) \sum_{k=0}^{[nx]} l(k,n) = R(x)$$

for any $x \in [0,q]$, $q = \min(1, C+\delta_0)$, where $R(x)$ is a non-decreasing function continuous at the $x=C$;

5) $g(n) \to 0$ as $n \to \infty$ and

$$\lim_{n\to\infty} \frac{g(nx)}{g(n)} = r(x), \quad x \in [0,C],$$

uniformly according to x from the set of the form $[a,b]$, $a>0$.

Theorem 3.3. If the conditions 1 - 5 are satisfied, then $g(n)X(n,\zeta(n))$ $\xrightarrow{d}$ X as $n\to\infty$, where

$$Me^{-\tau X} = \exp\left\{-\int_0^C \varphi\left(\frac{\tau}{r(1-x)}\right) r(1-x) dR(x)\right\}.$$

Remark 3.3. It is known (see Seneta, 1985, p. 17) that, if the limit in condition 5 exists, then $r(x) = x^\alpha$, $\alpha \in (0,\infty)$. Note also that conditions 3 and 5 can be satisfied, for example, for critical or almost critical branching processes with finite or infinite variance (see also §2.4).

Proof of Theorem 3.3. First we show that the conditions of Lemma 1.2.1 are satisfied for the sum from (3.27). We set

$$P_1 = P\left\{ g(n) \sum_{k=0}^{[m(n)M]} \max_{i \in \Delta_\delta(l(k,t))} \left| \sum_{j=1}^{i} \mu_{kj}(n-k) - \sum_{j=1}^{l(k,n)} \mu_{kj}(n-k) \right| > \varepsilon \right\},$$

where

$$\Delta_\delta(x) = \left\{ i \in N_0 : |i-x| < \delta x \right\}, \quad 0 < \varepsilon < \delta_0, \quad 1 < M < \frac{C+\delta_0}{C+\varepsilon}.$$

Let $\Delta_\delta^{(1)}(x)$ and $\Delta_\delta^{(2)}(x)$ be subsets of $\Delta_\delta(x)$ such that

§2.3. Discrete-Time Processes

$$i \in \Delta_\delta^{(1)}(x) \Leftrightarrow i \in \Delta_\delta(x), \ i \leq x; \quad i \in \Delta_\delta^{(2)}(x) \Leftrightarrow i \in \Delta_\delta(x), \ i > x.$$

It follows from Condition 2 that $m(n) \leq (C+\varepsilon)n$ for any sufficiently large n. Therefore, granting that $\mu_{ki}(n-k) = 0$ for $k > n$, and using Chebyshev's inequality, the inequality

$$\max_{i \in \Delta_\delta(x)} |f(i)| \leq \max_{i \in \Delta_\delta^{(1)}(x)} |f(i)| + \max_{i \in \Delta_\delta^{(2)}(x)} |f(i)| \qquad (3.28)$$

and the estimate of $M\mu_{ij}(n)$ from Condition 3 we obtain for any sufficiently large n that

$$P_1 \leq P\left\{ g(n) \sum_{k=0}^{[nq]} \sum_{j=[(1-\delta)l(k,n)]+2}^{[(1+\delta)l(k,n)]} \mu_{kj}(n-k) > \varepsilon \right\} \leq \frac{2C_0 \delta}{\varepsilon} g(n) \sum_{k=0}^{[nq]} l(k,n).$$

If we use condition 4 we have the relation

$$\limsup_{n \to \infty} P_1 \leq \frac{2C_0 \delta}{\varepsilon} R(q),$$

which shows that condition (II) of Lemma 1.2.1 is fulfilled.

Now we consider the probability

$$P_2 = P\left\{ g(n) \max_{i \in \Delta_\delta(m(n))} \left| \sum_{k=0}^{i} \sum_{j=1}^{l(k,n)} \mu_{kj}(n-k) - \sum_{k=0}^{m(n)} \sum_{j=1}^{l(k,n)} \mu_{kj}(n-k) \right| > \varepsilon \right\}.$$

Let $C < 1$. Then, it follows from Condition 4 that $|m(n) - nC| < \varepsilon_1 n$ for any sufficiently large n and any $\varepsilon_1 > 0$. Let $\varepsilon_1 < \delta_0$, $\delta < (\delta_0 - \varepsilon_1)(C + \varepsilon_1)^{-1}$. Using inequality (3.28) we have

$$P_2 = P\left\{ g(n) \left[\max_{i \in \Delta_\delta^{(1)}(m(n))} \sum_{k=i+1}^{m(n)} \sum_{j=1}^{l(k,n)} \mu_{kj}(n-k) \right. \right.$$

$$\left. \left. + \max_{i \in \Delta_\delta^{(2)}(m(n))} \sum_{k=m(n)+1}^{i} \sum_{j=1}^{l(k,n)} \mu_{kj}(n-k) \right] > \varepsilon \right\}$$

$$\leq \frac{C_0}{\varepsilon} g(n) \sum_{k=[(1-\sigma)m(n)]}^{[(1+\sigma)m(n)]} l(k,n). \qquad (3.29)$$

However the last sum is not greater than

$$\sum_{k=[q_1 n]}^{[q_2 n]} l(k,n), \quad q_1=(1-\delta)(C-\varepsilon_1), \quad q_2=(1+\delta)(C+\varepsilon_1),$$

where the $q_2 < C+\delta$ by virtue of the choice of ε_1 and δ. Using Condition 4 we obtain from (3.29) that

$$\limsup_{n\to\infty} P_2 \leq \frac{C_0}{\varepsilon} [R(q_2)-R(q_1)].$$

As the function $R(x)$ is continuous for $x=C$, it is possible to choose δ and ε_1 such that

$$\limsup_{n\to\infty} P_2 \leq \varepsilon. \qquad (3.30)$$

If $C=1$, using (3.28) we have:

$$P_2 = P\left\{ g(n) \sum_{k=[(1-\sigma)m(n)]}^{[(1+\sigma)m(n)]} l(k,n) \sum_{j=1}^{l(k,n)} \mu_{kj}(n-k) > \varepsilon \right\}.$$

Granting that $\mu_{kj}(n-k)=0$ when $k>n$, we obtain for any sufficiently large n and for any $\varepsilon_1 > 0$ that

$$\limsup_{n\to\infty} P_2 \leq \frac{C_0}{\varepsilon} [R(1)-R(\hat{q}_1)], \quad \hat{q}_1=(1-\delta)(1-\varepsilon_1).$$

If we choose a suitable δ and ε_1, then we can see that the relation (3.30) is true for $C=1$ also. Thus Condition (IV) of Lemma 1.2.1 is also fulfilled.

Thus, it follows from 1 - 4 that all the conditions of Lemma 1.2.1 are fulfilled for the sum from (3.27).

We now consider the process

$$\hat{X}(n) = \sum_{k=o}^{m(n)} \sum_{i=1}^{l(k,n)} \mu_{ki}(n-k).$$

If we set $f_n(s) = MS^{\mu_{ij}(n)}$, we have:

$$\ln MS^{\hat{X}(n)} = \sum_{k=o}^{m(n)} l(k,n) \ln f_{n-k}(s). \qquad (3.31)$$

Let $C<1$; without losing generality we can assume that $\delta_o < 1-C$. Choose $0 < \varepsilon_1 < \delta_o$. Then we obtain from condition 2 that $|m(n)-Cn| < \varepsilon_1 n$ for a

§2.3. Discrete-Time Processes

sufficiently large n and

$$I \equiv \sum_{k=0}^{m(n)} 1(k,n)(1-f_{n-k}(e^{-\tau g(n)})) \le \sum_{k=0}^{[n(C+\varepsilon_1)]} 1(k,n)(1-f_{n-k}(e^{-\tau g(n)})).$$

It follows from conditions 3 and 5 that for a sufficiently large n and $0 \le k \le [n(C+\varepsilon_1)]$

$$\left| \frac{g(n-k)}{g(n)} - r(1-\frac{k}{n}) \right| < \varepsilon_1, \qquad (3.32)$$

$$\left| \frac{1-f_{n-k}(e^{-\tau g(n-k)})}{g(n-k)} - \varphi(\tau) \right| < \varepsilon_1. \qquad (3.33)$$

Using (3.32) and (3.33) we obtain the estimate

$$I \le \sum_{k=0}^{[n(C+\varepsilon_1)]} \varphi\left(\frac{\tau}{r(1-k/n)-\varepsilon_1} \right) [r(1-k/n)]g(n)[T_n(k)-T_n(k-1)],$$

where $T_n(k) = \sum_{i=0}^{k} 1(i,n)$. Hence

$$\limsup_{n\to\infty} I \le \int_0^{C+\varepsilon_1} \varphi\left(\frac{\tau}{r(1-x)-\varepsilon_1} \right) [r(1-x)+\varepsilon_1]dR(x). \qquad (3.34)$$

By analogous arguments it is possible to show that

$$\limsup_{n\to\infty} I \ge \int_0^{C-\varepsilon_1} \varphi\left(\frac{\tau}{r(1-x)+\varepsilon_1} \right) [r(1-x)-\varepsilon_1]dR(x). \qquad (3.35)$$

From (3.34) and (3.35) we have:

$$\lim_{n\to\infty} I = \int_0^C \varphi\left(\frac{\tau}{r(1-x)} \right) r(1-x)dR(x). \qquad (3.36)$$

If C=1, granting that $\mu_{ki}(n-k)=0$ for k>n, we obtain

$$\sum_{k=0}^{[n(1-\varepsilon_1)]} 1(k,n)(1-f_{n-k}(e^{-\tau g(n)})) \le I \le$$

$$\le \sum_{k=0}^{n} 1(k,n)(1-f_{n-k}(e^{-\tau g(n)})). \qquad (3.37)$$

Using Condition 4 and estimation $\sup_n M\mu(n) \le C_0$ we have:

$$\limsup_{n\to\infty} \sum_{k=[n(1-\varepsilon_1)]}^{n} l(k,n)(1-f_{n-k}(e^{-\tau g(n)})) \le$$

$$\le \tau C_0 [R(1)-R(1-\varepsilon_1)].$$

By virtue of the continuity of $R(x)$ for $x=C$ it follows that the last sum in (3.37) tends, as $n\to\infty$, to the integral

$$\int_0^C \varphi\left(\frac{\tau}{r(1-x)}\right) r(1-x) dR(x).$$

Thus the relation (3.36) is true for all $C \in [0,1]$. In order to obtain the statement of the theorem, it is sufficient to use the fact that $\ln x \sim x-1$, $x \to 1$, Lemma 1.2.1 and relations (3.31) and (3.36). Theorem 3.3 is proved.

§2.4. CONVERGENCE TO JIRINA PROCESSES AND TRANSFER THEOREMS FOR THE BRANCHING PROCESS

§4a. The Model

We consider the family of random processes $\{Z^{(n)}(t), t \in [0,\infty), n \in \mathbb{N}\}$ defined as the following. Let there be given for each n:

a) a non-decreasing sequence of random variables $\{\theta_k^n, k \in \mathbb{N}\}$ such that:

$$P\left\{0 \le \theta_1^{(n)} \le \theta_2^{(n)} \le \ldots \le \lim_{k\to\infty} \theta_k^{(n)} = \infty\right\} = 1; \qquad (4.1)$$

b) a family of independent and identically distributed random processes $\{\zeta_k^{(n)}(t), t \in [0,\infty), k \in \mathbb{N}\}$ taking non-negative values and independent of the family $\{\theta_k^n\}$.

We put

$$Z^{(n)}(t) = \sum_{k: \theta_k^{(n)} \le t} \zeta_k^{(n)}(t-\theta_k^{(n)}), \qquad \xi^{(n)}(t) = \sum_{k: \theta_k^{(n)} \le t} 1, \quad t \ge 0.$$

Such a family of processes was studied by Badalbaev and Zubkov (1983) under the following conditions:

$$\sup_{n \ge 0} \sup_{0 \le t \le \varepsilon_1 n} M\zeta_k^{(n)}(t) \le M_0, \qquad (4.2)$$

for some $\varepsilon_1 > 0$ and

§2.4. The Convergence to Jirina Processes

$$F_n(t,s) \equiv MS^{\zeta_k^{(n)}(t)} = 1 - \frac{1-s}{1+(1-s)t\gamma}(1+\alpha_n(t,s)) \qquad (4.3)$$

for some $\gamma \in (0,\infty)$, where $\alpha_n(t,s) \to 0$ as $n \to \infty$ uniformly in any domain of the form $\{\varepsilon n \leq t \leq n, |s| \leq 1\}$, $\varepsilon > 0$.

In this section we shall introduce a more natural assumption on processes $\zeta_k^{(n)}(t)$. More exactly, we shall suppose the existence of a weak limit of normalized and conditional processes $\{\zeta_k^{(n)}(t) | \zeta_k^{(n)}(t) \neq 0\}$. This circumstance allows us to obtain a result from which it is possible to deduce limit theorems for branching processes with immigration if a limit theorem for corresponding branching process without immigration is given. Above we called such results "transfer theorems for branching processes" (see §2.1).

We introduce the following conditions: let for any $\lambda > 0$

$$\frac{1-F_n(t,\exp\{-\lambda Q^{(n)}(t)\})}{Q^{(n)}(t)} \to 1 - \varphi(\lambda) \qquad (4.4)$$

as $n \to \infty$ uniformly in any domain of the form $\{\varepsilon n \Delta \leq t \leq n\Delta\}$, $\varepsilon > 0$, $\Delta > 0$, where $\varphi(\lambda)$ is the Laplace transform of a random variable with finite expectation, with $Q^{(n)}(t)$ being a function non-increasing according to t for any n, such that $Q^{(n)}(0) > 0$, $Q^{(n)}(n) \to 0$, $n \to \infty$, and for any $x \in [0,a]$

$$Q^{(n)}(n)/Q^{(n)}(xn) \to \pi(x), \qquad (4.5)$$

as $n \to \infty$ uniformly in each interval of the form $[\varepsilon, a]$, where $\varepsilon > 0$, and $\pi(x)$ is a continuous function for $x \in (0, a]$.

We now define a branching process with continuous state space and non-homogeneous immigration outstanding as a limit process for $Z^{(n)}(t)$. Let $P_t(x,y)$ be a family of functions such that:

1) $P_t(x,E)$ is defined for $t \geq 0$, $x \geq 0$ and for each Borel set on $[0, \infty)$.

2) For any fixed t and x the function $P_t(x, \cdot)$ is a probability measure on Borel sets and for any fixed E is measurably with respect to t and x.

3) Satisfies the following Kolmogorov-Chapman equation:

$$\int_0^\infty P_t(u,E) P_s(x,du) = P_{t+s}(X,E).$$

4) For any x, y, $t \geq 0$

$$P_t(x+y, \cdot) = P_t(x, \cdot) * P_t(y, \cdot).$$

5) There exist $t > 0$, $x > 0$ such that $P_t(x, \{0\}) < 1$.

II. Branching Processes with Generalized Immigration

Definition 4.1. [Jirina (1958), Lamperti (1967b)]. The Markov process $\zeta(t)$, $t\in[0,\infty)$, with transient probabilities

$$P_t(x,E) = P\{\zeta(s+t)\in E|\zeta(s)=x\}$$

satisfying conditions 1 - 5 is called a branching process with continuous state space or a Jirina process.

It follows from condition 4 that the distribution $P_t(x,\cdot)$ is infinitely divisible, therefore, its Laplace transform is representable in the form

$$\int_0^\infty e^{-\lambda y} P_t(x,dy) = e^{-x f_t(\lambda)}, \quad f_t(\lambda)\geq 0, \; \lambda\geq 0. \tag{4.6}$$

Using (4.6) we can write condition 3 in the form

$$f_t(f_s(\lambda)) = f_{t+s}(\lambda). \tag{4.7}$$

Let $T(x)$, $x\in[0,\infty)$, be a random process with non-decreasing trajectories and taking non-negative values.

Definition 4.2. The branching process with continuous state space, and immigration defined by the process $T(x)$, is a process $Y(t)$ whose finite-dimensional distributions have the Laplace transform

$$M\exp\left\{-\sum_{i=1}^r \lambda_i Y(t_i)\right\} = M e^{-K_r(t_1,\ldots,t_r,\lambda_1,\ldots,\lambda_r)},$$

where

$$r=1,2,\ldots, t_i < t_{i+1}, \quad K_1(t_1,\lambda_1) = \psi(0,t_1,\lambda_1),$$

$$K_j(t_1,\ldots,t_j,\lambda_1,\ldots,\lambda_j) = \psi(t_{j-1},t_j,\lambda_j) +$$

$$+ K_{j-1}(t_1,\ldots,t_{j-1},\lambda_1,\ldots,\lambda_{j-2},\lambda_{j-1}+f_{t_j-t_{j-1}}(\lambda_j)),$$

$$\psi(s,t,\lambda) = \int_s^t f_{t-x}(\lambda)dT(x).$$

§4b. The Main Theorem and Corollaries

Let $Y(t)$ be a branching process with continuous state space and immigration defined by process $T(x)$ and $f_t(\lambda) = (1-\varphi(\lambda\pi(t)))/\pi(t)$. We put

$$Y^{(n)}(t) = Q^{(n)}(n) Z^{(n)}(nt), \quad t\in[0,a], \; a\in(0,\infty).$$

§2.4. The Convergence to Jirina Processes

Theorem 4.1. Let $\{\varphi_k^{(n)}(t), k \geq 1\}$ be a family of independent and identically distributed Markov branching processes, and conditions (4.1), (4.2), (4.4), (4.5), (4.7) be fulfilled and finite-dimensional distributions of the process $\tau^{(n)}(x) = Q^{(n)}(n)\xi^{(n)}(nx)$, $x \in [0,a]$, tend to ones of the process $T(x)$, $P\{T(a) < \infty\} = 1$. If the process $T(x)$ is stochastically continuous at the points $x = t_i$, $t_i < t_{i+1}$, $t_i \in [0,a]$, $i = 1,\ldots,r$, then distribution of the vector $(Y^{(n)}(t_1), \ldots, Y^{(n)}(t_r))$ tends to the distribution of $(Y(t_1),\ldots,Y(t_r))$.

Corollary 4.1. If under the conditions of Theorem 4.1 the process $T(x)$ is stochastically continuous on $[0,a]$, then $Y^n(t) \to Y(t)$, $t \in [0,a]$, as $n \to \infty$ in the meaning of the finite-dimensional distributions.

The following lemma is subsidiary to proof of Theorem 4.1 and is also of an independent interest.

Lemma 4.1. Let the conditions (4.1), (4.2), (4.4) and (4.5) be satisfied and the finite-dimensional distributions of the process $\tau^{(n)}(x)$, $x \in [0,a]$, tend as $n \to \infty$ to ones of the process $T(x)$, $x \in [0,a]$, $P\{T(a) < \infty\} = 1$, with non-decreasing trajectories and which is stochastically continuous for $x = a$. Then:

$$\lim_{n \to \infty} M \exp\{-\lambda Y^{(n)}(a)\} = M \exp\left\{-\int_0^a f_{a-x}(\lambda) dT(x)\right\}$$

(the value of the integrand at $x = a$ is defined by continuity).

Note that we do not assume here, in contrast to Theorem 4.1, that $\zeta_k^{(n)}(t)$ are branching processes.

Corollary 4.2. If, under the conditions of Lemma 4.1, the process $T(x)$ is stochastically continuous on $[0,a]$, then $Y^{(n)}(t) \xrightarrow{d} Y(t)$, $t \in [0,a]$, where

$$Me^{i\lambda Y(t)} = M \exp\left\{-\int_0^t f_{t-x}(\lambda) dT(x)\right\}.$$

Now we shall present some corollaries of Theorem 4.1 and Lemma 4.1.

1°. First we note that, if $a = 1$ and the condition (4.3) is fulfilled, then we can find from Lemma 4.1 a result obtained by Badalbaev and Zubkov (1983).

2°. Let $\{\zeta_k^{(n)}(t), k \geq 1\}$ be a family of general Krump-Mode-Jagers processes, let μ be the number of direct offspring with

$$MS^\mu = s + (1-s)^{1+\nu} L(1-s), \quad \nu \in (0,1], \quad (4.8)$$

where $L(x)$ is a slowly varying function as $x \to 0$, and let $G(t)$ be the distribution of lifetime of particles, $A(t)$ be the expectation of the offspring number of one particle at time t,

$$\beta = \int_0^\infty u\,dG(u) \Big/ \int_0^\infty u\,dA(u).$$

We shall use the following statement from the paper of Sagitov (1983).

Proposition 4.1. If (4.8) is satisfied, $A(t)$ is a non-lattice distribution function, $\beta \in (0,\infty)$ and

$$t^{1+\nu}(1-G(t))^\nu L((1-G(t))t) \to 0,$$

$$t^{1+\nu}(1-A(t))^\nu L((1-A(t))t) \to 0, \quad t\to\infty,$$

then the conditions (4.2), (4.4) and (4.5) are satisfied for the process $\{\varphi_k^{(n)}(t)\}$ with

$$Q(t)=P\{\zeta_k^{(n)}(t)>0\}, \quad \varphi(\lambda)=1-(1+(\beta\lambda)^{-\nu})^{-1/\nu}, \quad \pi(x)=x^{1/\nu}.$$

Under the conditions of Proposition 4.1 it is possible to obtain corresponding limit theorems for general branching processes with immigration and with infinite variance from Lemma 4.1.

3°. Let $\zeta^{(n)}(t)\equiv\zeta(t)$, $t\in N_0$, be a Galton-Watson process, $Q(t)=P(\zeta(t)>0)$ and let the immigration process be defined by the sequence of independent random variables $\{I_k^{(n)}, k=0,1,\ldots,n\}$. Assume that

$$Q(n)\sum_{k=0}^n I_k^{(n)} \xrightarrow{D} I, \quad Me^{-\lambda I}=\exp\left\{-\int_0^\infty \frac{1-e^{-\lambda x}}{x} P(dx)\right\}. \tag{4.9}$$

Corollary 4.3. Under conditions (4.8), (4.9)

$$Q(n)Z^{(n)}(n) \xrightarrow{D} Z_\nu, \quad n\to\infty,$$

where

$$Me^{-\lambda Z_\nu}=\exp\left\{-\int_0^\infty \frac{1-e^{-\lambda x}}{x} Q_\nu(dx)\right\}, \tag{4.10}$$

$Q_\nu = P*M_\nu$, P is defined by (4.9) and M_ν is defined by the relation

$$\int_0^\infty e^{-\lambda x} M_\nu(dx) = (1+\lambda^\nu)^{-1}.$$

Corollary 4.3 is an extended version of Foster and Williamson's theorem (1971).

4°. As before, let $\zeta(n)$, $n\in N_0$, be a Galton-Watson process satisfying

§2.4. The Convergence to Jirina Processes

condition (4.8). Consider the immigration process defined by the family of random variables $(I_k^{(n)}, k=0,1,\ldots,n)$, where $I_k^{(n)} \equiv I_k$, $k \geq 1$, and

$$I_0^{(n)} Q(n) \xrightarrow{D} c, \quad Q(n) \sum_{k=1}^{n} I_k \xrightarrow{D} d, \quad n \to \infty,$$

where c and d are some constants. Then it is not difficult to see that

$$Q(n)\xi^{(n)}(nx) \xrightarrow{D} c + x^{1/\nu} d$$

as $n \to \infty$, that is, $T(x) = c + x^{1/\nu} d$ for $0 \leq x < \infty$ and $T(0-) = 0$. Simple calculations show that in this case

$$Me^{-\lambda Y(t)} = \exp\left\{ - \frac{c}{(t+\lambda^{-\nu})^{1/\nu}} - \frac{d}{\nu} \int_0^t \frac{x^{1/\nu - 1} dx}{(t-x+\lambda^{-\nu})^{1/\nu}} \right\}, \quad \lambda \geq 0, \ t \geq 0.$$

If $\nu = 1$, then

$$Me^{-\lambda Y(t)} = (1+t\lambda)^{-d} \exp\left\{ - \frac{\lambda c}{1+t\lambda} \right\}. \tag{4.11}$$

It is known that the process $Y(t)$ with the Laplace transform of the form (4.11) is a homogeneous diffusion and its density of transient probabilities $U(t,c)$ satisfies Kolmogorov's differential equation

$$\frac{\partial u}{\partial t} = d \frac{u}{c} + c^2 \frac{\partial^2 u}{\partial c^2},$$

with initial condition $U(0,c) = e^{-\lambda c}$ (see Lamperti, 1967b).

§4c. Proof of the Main Theorem

Proof of Lemma 4.1. Let $0 < \varepsilon < 1$, $E = 1-\varepsilon$,

$$Z_1^{(n)}(\varepsilon, t) = \sum_{k: 0 < \theta_k^{(n)} \leq tE} \zeta_k^{(n)}(t - \theta_k^{(n)}),$$

$$Z_2^{(n)}(\varepsilon, t) = \sum_{k: tE < \theta_k^{(n)} \leq t} \zeta_k^{(n)}(t - \theta_k^{(n)}).$$

Using the independence of processes $\{\zeta_k^n(t)\}$ and random variables $\{\theta_k^{(n)}\}$ we obtain the relation

$$\text{MS} \, Z_1^{(n)}(\varepsilon,t) = M \exp\left\{\int_0^{tE} \ln F_n(t-u,s) d\xi^{(n)}(u)\right\}. \tag{4.12}$$

By condition (4.5) for sufficiently large n and $0 \leq u \leq ntE$ we have:

$$0 < \pi(\varepsilon t) - \varepsilon \leq \pi_n(t-u/n) \leq \pi(t) + \varepsilon, \tag{4.13}$$

where $\pi_n(x) = Q^{(n)}(n)/Q^{(n)}(nx)$. Then using (3.13) we obtain from (4.4) that as $n \to \infty$

$$\sup_{0 \leq u \leq tnE} \left| \frac{1 - F_n(nt-u, e^{-\lambda Q^{(n)}(n)})}{Q^{(n)}(nt-u)} - 1 + \varphi(\lambda \pi(t-u/n)) \right| \to 0. \tag{4.14}$$

Since $0 < t \varepsilon \leq t - u/n \leq t < \infty$ when $0 \leq u \leq ntE$, it follows from (4.5) that

$$\sup_{0 \leq u \leq tnE} \left| \frac{Q^{(n)}(nt-u)}{Q^{(n)}(n)} - \frac{1}{\pi(t-u/n)} \right| \to 0, \quad n \to \infty. \tag{4.15}$$

Thus from (4.14), (4.15) and condition (4.5) we obtain the following representation:

$$-\frac{1}{Q^{(n)}(n)} \ln F_n(nt-u, e^{-\lambda Q^{(n)}(n)}) = f_{t-u/n}(\lambda)(1 + \hat{\alpha}_1(n,u)), \tag{4.16}$$

where $\hat{\alpha}_1(n,u) \to 0$ as $n \to \infty$ uniformly with respect to $u \in [0, ntE]$.

Since $\tau^{(n)}(x) \to T(x)$ as $n \to \infty$ in the meaning of finite-dimensional distributions,

$$\int_0^{ntE} f_{t-u/n}(\lambda) d(Q^{(n)}(n)\xi^{(n)}(u))$$

$$= \int_0^{tE} f_{t-x}(\lambda) d\tau^{(n)}(x) \to \int_0^{tE} f_{t-x}(\lambda) dT(x). \tag{4.17}$$

On the other hand,

$$\left| \int_0^{tE} f_{t-x}(\lambda) dT(x) - \int_0^{t} f_{t-x}(\lambda) dT(x) \right| \leq -\lambda \varphi'(0)[T(t) - T(tE)] \tag{4.18}$$

and thus the expression on the right-hand side of (4.18) converges to zero in probability when $\varepsilon \to 0$ in view of the stochastic continuity of $T(x)$.

§2.4. The Convergence to Jirina Processes

It is possible to find $\varepsilon \in (0, \varepsilon_1/t)$ and $n_0(\varepsilon) < \infty$ such that for any $\delta > 0$ and $n > n_0(\varepsilon)$

$$P\{Q^{(n)}(n) Z_2^{(n)}(\varepsilon, nt) > \delta\} < \delta. \qquad (4.19)$$

From conditions (4.17)-(4.19) and relation $Z = Z_1 + Z_2$ we obtain the statement of Lemma 4.1. We set

$$\Lambda_\varepsilon^{(n)}(t_1, t_2, x) = \int_{t_1 E}^{t_2 E} \ln F_n(t_2 - u, x) d\xi^{(n)}(u).$$

We define the functions

$$h_r^{(n)}(\varepsilon, t, s) = h_r^{(n)}(\varepsilon, t_1, \ldots, t_r, s_1, \ldots, s_r)$$

by the relations

$$h_1^{(n)}(\varepsilon, t, s) = \Lambda_\varepsilon^{(n)}(0, t_1, s_1),$$

$$h_r^{(n)}(\varepsilon, t, s) = \Lambda_\varepsilon^{(n)}(t_{r-1}, t_r, s_r) + h_{r-1}^{(n)}(\varepsilon, t, \hat{s}),$$

$$\hat{s} = (s_1, \ldots, s_{r-2}, \hat{s}_{r-1}), \quad \hat{s}_{r-1} = s_{r-1} F_n(t_r - t_{r-1}, s_r).$$

Note that here and later on $t = (t_1, \ldots, t_r)$, $s = (s_1, \ldots, s_r)$ are vectors with dimensions defined by the numbers of corresponding functions. For example, $s = (s_1, s_2, s_3)$ in $h_3(s)$.

Lemma 4.2. For any t_1 and $0 < \varepsilon < 1$

$$M s_1^{Z_1^{(n)}(\varepsilon, t)} = M \exp\{\Lambda_\varepsilon^{(n)}(0, t_1, s_1)\}$$

and if, $\{\zeta_k^{(n)}(t), k \in N\}$ are independent and identically distributed branching processes, then for the generating function

$$\Phi_r^{(n)}(\varepsilon, t, s) = M\left[\prod_{i=1}^{r} s_i^{Z_1^{(n)}(\varepsilon, t_i)}\right], \quad r \geq 1$$

the following relation is true

$$\Phi_r^{(n)}(\varepsilon, t, s) = M \exp\{h_r^{(n)}(\varepsilon, t, s)\}.$$

The proof of this lemma follows from the definition of $Z(t)$ by the total probability formula.

Let $\lambda=(\lambda_1,\ldots,\lambda_r)$, $e^{-\lambda}=(e^{-\lambda_1},\ldots,e^{-\lambda_r})$ and functions $K_r(\varepsilon,t,\lambda)$ are defined by the relations:

$$K_1(\varepsilon,t,\lambda) = \psi_\varepsilon(0,t_1,\lambda_1),$$

$$K_r(\varepsilon,t,\lambda) = \psi_\varepsilon(t_{r-1},t_r,\lambda_r)+K_{r-1}(\varepsilon,t,\hat\lambda),$$

where

$$\hat\lambda=(\lambda_1,\ldots,\lambda_{r-2},\hat\lambda_{r-1}), \quad \hat\lambda_{r-1}=\lambda_{r-1}+f_{t_r-t_{r-1}}(\lambda_r),$$

$$\psi_3(t_1,t_2,\lambda_1) = \int_{t_1 E}^{t_2 E} f_{t_2-x}(\lambda_1)dT(x),$$

and the function $f_t(\lambda)$ is defined in Theorem 4.1.

Lemma 4.3. Let $h_r^{(n)}(\varepsilon,t,s)$, $r\geq 1$, be the functions defined in Lemma 4.2. If (4.1), (4.4) and (4.5) are satisfied, and the finite-dimensional distributio of $\tau^{(n)}(x)$ converge to ones of the process $T(x)$ which is stochastically continuous at $x=t_i$, with $t_i\in[0,a]$, $i=1,\ldots,r$, then

$$h_r^{(n)}(\varepsilon,nt,e^{-\lambda Q^{(n)}(n)}) \xrightarrow{D} K_r(\varepsilon,t,\lambda), \quad n\to\infty.$$

Proof. The statement of the lemma in the case $r=1$ follows from lemma 4.1. Let now $r\geq 2$. It is not difficult to see that

$$h_r(\varepsilon,t,s)= \sum_{i=1}^{r} \Lambda_\varepsilon(t_{i-1},t_i,\hat S_i), \quad K_r(\varepsilon,t,\lambda)= \sum_{i=1}^{r} \psi_\varepsilon(t_{i-1},t_i,\hat\lambda_i), \tag{4.20}$$

where

$$\hat S_i=S_i F(t_{i+1}-t,\hat S_{i+1}), \quad \hat\lambda_i=\Lambda_i+f_{t_{i+1}-t_i}(\hat\lambda_{i+1}),$$

$$i=r-1,r-2,\ldots,2,1, \quad t_0=0, \quad \hat S_r=S_r, \quad \hat\lambda_r=\lambda_r.$$

The same arguments as in the proof of relation (4.16) show that

$$-\frac{\ln F_n\left(nt_i-u, e^{-\lambda_i Q^{(n)}(n)}\right)}{Q^{(n)}(n)} = f_{t_i-u/n}(\lambda_i)(1+\alpha(n.u)), \tag{4.21}$$

where $\alpha(n,u)\to 0$ uniformly on $u\in[nt_{i-1}E, nt_i E]$ as $n\to\infty$.

§2.4. The Convergence to Jirina Processes

It is obvious that under conditions (4.4) and (4.5)

$$\lim_{n\to\infty} \frac{\ln F_n\left(n(t_i-t_{i-1}), e^{-\lambda_i Q^{(n)}(n)}\right)}{Q^{(n)}(n)} = -f_{t_i-t_{i-1}}(\hat{\lambda}_i),$$

for any $t_{i-1} < t_i$, that is,

$$e^{-\lambda_{i-1} Q^{(n)}(n)} F_n\left(n(t_i-t_{i-1}), e^{-\hat{\lambda}_i Q^{(n)}(n)}\right) = e^{-\hat{\lambda}_{i-1}^{(n)} Q^{(n)}(n)},$$

where $\lambda_{i-1}^{(n)} \to \lambda_{i-1} + f_{t_i-t_{i-1}}(\hat{\lambda}_i) \overset{\text{def}}{=} \hat{\lambda}_{i-1}$, $n \to \infty$.

Then, for a sufficiently large n and for any $\varepsilon > 0$ and $i = 2, 3, \ldots, r$,

$$e^{-(\hat{\lambda}_{i-1}+\varepsilon_0)Q^{(n)}(n)} \le e^{-\hat{\lambda}_{i-1}^{(n)} Q^{(n)}(n)} \le e^{-(\hat{\lambda}_{i-1}-\varepsilon_0)Q^{(n)}(n)}.$$

Hence, granting the monotonicity of $F_n(t,s)$ according to s and relation (4.20), we obtain the inequality:

$$-h_r(\varepsilon, nt, e^{-(\hat{\lambda}-\varepsilon_0)Q^{(n)}(n)}) \le -h_r(\varepsilon, nt, e^{-\lambda Q^{(n)}(n)}) \le$$
$$\le -h_r(\varepsilon, nt, e^{-(\hat{\lambda}+\varepsilon_0)Q^{(n)}(n)}), \qquad (4.22)$$

where $\hat{\lambda} \pm \varepsilon_0 = (\hat{\lambda}_1 \pm \varepsilon_0, \ldots, \hat{\lambda}_r \pm \varepsilon_0)$. If we denote by $P_{+\varepsilon_0}(n,x)$, $P(n,x)$ and $P_{-\varepsilon_0}(n,x)$ the distributions of variables from (4.22) respectively, then

$$P_{+\varepsilon_0}(n,x) \le P(n,x) \le P_{-\varepsilon_0}(n,x). \qquad (4.23)$$

Using relations (4.20) and (4.21) we obtain that

$$-h_r(\varepsilon, nt, e^{-(\hat{\lambda}\pm\varepsilon_0)Q^{(n)}(n)}) = o(Q^{(n)}(n)\xi^{(n)}(nt_r E)) +$$
$$+ \sum_{i=1}^{r} \int_{nt_{i-1}E}^{nt_i E} f_{t_i-u/n}(\hat{\lambda}_i \pm \varepsilon_0) d(\xi^{(n)}(u)Q^{(n)}(n)). \qquad (4.24)$$

After a change of variables, the second term will have the following form:

$$\sum_{i=1}^{r} \int_{t_{i-1}E}^{t_i E} f_{t_i - x}(\hat{\lambda}_i \stackrel{+}{-} \varepsilon_0) d\tau_n(x). \qquad (4.25)$$

Since $\tau_n(x) \to T(x)$ as $n \to \infty$ in the meaning of the finite-dimensional distributions, the distribution of (4.25) tends to the distribution of the variable

$$\sum_{i=1}^{r} \int_{t_{i-1}E}^{t_i E} f_{t_i - x}(\hat{\lambda}_i \stackrel{+}{-} \varepsilon_0) dT(x). \qquad (4.26)$$

If we denote by $R_{\stackrel{+}{-}\varepsilon_0}(x)$ the distribution of (4.26), then by (4.23) we have

$$R_{+\varepsilon_0}(x) \leq \liminf_{n \to \infty} P(n,x) \leq \limsup_{n \to \infty} P(n,x) \leq R_{-\varepsilon_0}(x). \qquad (4.27)$$

Since $|\varphi'(0)| < \infty$, random variable (4.26) converges to

$$\sum_{i=1}^{r} \int_{t_{i-1}E}^{t_i E} f_{t_i - x}(\hat{\lambda}_i) dT(x) = K_r(\varepsilon, t, \lambda)$$

in probability when $\varepsilon_0 \to 0$.

Thus we obtain from (4.27) the statement of Lemma 4.3 for $r \geq 2$. The lemma is thus proved.

Lemma 4.4. Let conditions (4.1) and (4.2) be satisfied, $\tau^{(n)} \to T(x)$, $x \in [0,a]$, as $n \to \infty$ in the meaning of finite-dimensional distributions, where $T(x)$ is stochastically continuous at $x = x_i$, $t_i < t_{i+1}$, $t_i \in [0,a]$, $i = 1, 2, \ldots, r$. Then there are $\varepsilon \in (0, \varepsilon_0/t_r)$ and $n_0(\varepsilon) < \infty$ such that for any $\delta > 0$, $\lambda > 0$ and $n > n_0(\varepsilon)$

$$P\left\{Q^{(n)}(n) \sum_{i=1}^{r} \lambda_i Z_2^{(n)}(\varepsilon, nt_i) > \delta\right\} < \delta.$$

Proof. Let

$$A_n(\delta) = \left\{Q^{(n)}(n) \sum_{i=1}^{r} \lambda_i Z_2^{(n)}(\varepsilon, nt_i) > \delta\right\},$$

$$B_n(\delta_1) = \left\{Q^{(n)}(n) \sum_{i=1}^{r} \lambda_i (\xi^{(n)}(nt_i) - \xi^{(n)}(nt_i E)) > \delta_1\right\}.$$

It is obvious that

$$P\{A_n(\delta)\} = P\{A_n(\delta) \cap B_n(\delta_1)\} + P\{A_n(\delta) \cap \overline{B_n(\delta_1)}\}, \qquad (4.28)$$

where

§2.4. The Convergence to Jirina Processes

$$P\{A_n(\delta) \cap B_n(\delta_1)\} \leq$$

$$\leq \sum_{i=1}^{r} P\left\{\lambda_i(\xi^{(n)}(nt_i) - \xi^{(n)}(nt_i E))Q^{(n)}(n) > \delta_1/r\right\}. \qquad (4.29)$$

Under the conditions of the lemma, the sum on the right-hand side of (4.29) tends to the sum

$$\sum_{i=1}^{r} P\left\{\lambda_i(T(t_i) - T(t_i E)) > \delta_1/r\right\}.$$

Each term of the last sum tends to zero as $\varepsilon \to 0$ for fixed δ_1 and r, in view of the stochastic continuity of $T(x)$ at points $t_i, i=1,\ldots,r$. Thus, for any sufficiently large n and small ε,

$$P\{A_n(\delta) \cap B_n(\delta_1)\} < \delta/2 . \qquad (4.30)$$

Using Chebyshev's inequality we obtain that

$$P\{A_n(\delta) \cap \bar{B}_n(\delta_1)\} \leq \frac{1}{\delta} M\left[\sum_{i=1}^{r} \lambda_i z_2^{(n)}(\varepsilon, nt_i) Q^{(n)}(n) \chi(\bar{B}_n(\delta_1))\right]. \qquad (4.31)$$

It is easy to show that the expectation on the right-hand side of (4.31) is not greater than $\delta^2/2$. Hence

$$P\{A_n(\delta) \cap \bar{B}_n(\delta_1)\} < \delta/2. \qquad (4.32)$$

The statement of the lemma follows from (4.28) and (4.32).

Now we prove the basic theorem.

Proof of Theorem 4.1. It follows from Lemma 4.2 that

$$\Phi_r^{(n)}(\varepsilon, t, e^{-\lambda}) = M \exp\{h_r^{(n)}(\varepsilon, t, e^{-\lambda})\}.$$

Using Lemma 4.3 we obtain:

$$\lim_{n \to \infty} \Phi_r^{(n)}(\varepsilon, nt, e^{-\lambda Q^{(n)}(n)}) = M e^{-K_r(\varepsilon, t, \lambda)}. \qquad (4.33)$$

If we use the relation (4.20), we have:

$$|K_r(\varepsilon, t, \lambda) - K_r(0, t, \lambda)| \leq$$

$$\leq -\varphi'(0) \sum_{i=1}^{r} \hat{\lambda}_i [T(t_i) + T(t_{i-1}) - T(t_i E) - T(t_{i-1} E)],$$

where the sum on the right-hand side converges to zero in probability as $\varepsilon \to 0$. Therefore, by (4.33) we obtain that there are $\varepsilon \in (0, \varepsilon_1/t_1)$ and $n_o(\varepsilon) < \infty$ such that

II. Branching Processes with Generalized Immigration

$$|\Phi_r^{(n)}(\varepsilon, nt, e^{-\lambda Q^{(n)}(n)}) - Me^{-K_r(0,t,\lambda)}| < \delta \qquad (4.34)$$

for any $\delta > 0$, $\lambda > 0$ and $n \geq n_o$.

Using Lemma 4.4 we obtain the estimate

$$0 \leq Me^{-\sum_{i=1}^{r} \lambda_i z_1(\varepsilon, nt_i) Q^{(n)}(n)} - Me^{-\sum_{i=1}^{r} \lambda_i z(nt_i) Q^{(n)}(n)} \leq 2\delta.$$

The statement of Theorem 4.1 follows from (4.34) and the last inequality. The theorem is now proved.

CHAPTER III

BRANCHING PROCESSES WITH TIME-DEPENDENT IMMIGRATION

In this chapter we will investigate in detail the asymptotic behavior of a concrete model of branching processes with immigration dependent on time. More exactly, we will consider the Galton-Watson process with non-homogeneous (independent of reproduction) immigration. In contrast to Chapter II, we shall demonstrate here a combination of analytical and probability methods.

The asymptotic behavior of non-homogeneous processes depends strongly on the behavior of immigration parameters at infinity. Therefore, it is natural to split the set of all processes into classes, depending on their limit behavior.

Definition 0.1. We say that the process $Z(t)$ belongs to the class $A(\varphi_t(y), \pi(t,x))$ if for any $x \in \Delta_2 \subseteq R$, $\Delta_2 \neq \emptyset$

$$\lim_{t \to \infty} \frac{P\{\varphi_t(Z(t)) < x \mid Z(t) > 0\}}{\pi(t,x)} = 1, \qquad (0.1)$$

and belongs to the class $B(\varphi_t(y), \pi(t,x))$, if relation (0,1) is true for the unconditional distributions. We view the set Δ_1 either as $N_o = \{0,1,2,\ldots\}$ or as $R_+ = [0,\infty)$ and the set Δ_2 will be described under the definition of the function $\pi(t,x)$.

§3.1. DECREASING IMMIGRATION

Let $\mu(n)$, $n \in N_o$, $\mu(0)=1$, be the Galton-Watson process, let $f(s)$, $0 \leq s \leq 1$, be the generating function of the distribution of the number of direct offspring of one particle. Assume a random number ξ_n of new particles immigrate at time n. Each of these particles, independently of others, undergo transformations according to the process $\mu(n)$. We also assume that the random variables ξ_n, $n \in N_o$, are independent of each other and of the process $\mu(n)$. Denote by $Z(n)$, $Z(0)=0$, the process with immigration. It is clear that in this case the immigration process is defined as the sum of independent random variables $X_n = \xi_1 + \xi_2 + \cdots + \xi_n$.

§1a. The Main Theorem

In this section we prove limit theorems for $Z(n)$ when $M\xi_n \to 0$, $n \to \infty$, under the assumption that $f(s)$ is a regularly varying function. It turned out that if the series $\sum_n P\{\mu(n)>0\}$ is divergent, then the limit distributions of $Z(n)$ are the same for process with finite variance. If this series converges we obtain some new limit distributions.

Here and later on we assume that the function $f(s)$ is representable in the form:

$$f(s) = s+(1-s)^{1+\nu}L(1-s), \qquad (1.1)$$

where $0<\nu\le 1$, and $L(x)$ is a slowly varying function as $x\to 0$.

As shown in Harris (1966), if $0<f(0)<1$, then there exists a stationary measure of the process $\mu(n)$. The generating function of this stationary measure $U(s)$ is analytic for $|s|<q$ (q is the extinction probability) and, under the condition $U(f(0))=1$, satisfies the following equation:

$$U(f(s))=1+U(s), \quad |s|<q, \quad U(1)=\infty. \qquad (1.2)$$

In addition if $f'(0)\ne 1$, then Green's function of the process $\mu(n)$

$$G(i,j) = \sum_{n=0}^{\infty} P\{\mu(n)=j \mid \mu(0)=i\}$$

is finite for any $i\in N_0$, $j\in N$, (see Athrea and Ney (1972)).

As proved by Slack (1968), under condition (1.1)

$$U(s) = \frac{(1-s)(1+o(1))}{\nu(f(s)-s)} = \frac{1+o(1)}{\nu(1-s)^{\nu}L(1-s)}, \quad s \nearrow 1. \qquad (1.3)$$

It follows from (1.3) that the inverse function of $U(1-s)$ has the following form:

$$g(x)=N(x)x^{-1/\nu}, \quad x>0, \qquad (1.4)$$

where $N(x)$ is a slowly varying function as $x\to\infty$ and

$$N^{\nu}(x)L(N(x)/x^{1/\nu}) \to \nu^{-1}. \qquad (1.5)$$

Let

§3.1. The Decreasing Immigration

$$q_k(n) = P\{\xi_n = k\}, \quad h_n(s) = \sum_{k=0}^{\infty} q_k(n) s^k,$$

$$\alpha(n) = M\xi_n, \quad \beta(n) = D\xi_n + \alpha^2(n) - \alpha(n).$$

Since the random variables ξ_n are independent,

$$F_n(s) = MS^{Z(n)} = \prod_{k=0}^{n} h_k(f_{n-k}(s)), \tag{1.6}$$

where $f_n(s) = f(f_{n-1}(s))$, $f_o(s) = s$. Pakes (1975) shows that (see also Pakes, 1976), if $f'(1) = 1$, then

$$1 - f_n(s) = g(n + U(s)). \tag{1.7}$$

In this section we assume that $\alpha(n) \to 0$ and is regularly varying function as $n \to \infty$ and

$$\sup_{0 \le k < \infty} \alpha(k) < \infty, \quad \sup_{0 \le k < \infty} \beta(k) < \infty. \tag{1.8}$$

Under condition (1.8) (see Sevast'yanov, 1971, p. 18) for any k

$$h_k(s) = 1 + \alpha(k)(s-1) + 2^{-1}\hat{\beta}_k(s)(1-s)^2, \tag{1.9}$$

where $0 \le \hat{\beta}_k(s) \le \beta(k)$, when $0 \le s \le 1$.

Introduce the following notations:

$$a_1(n) = \sum_{k=0}^{n} P\{\mu(k) > 0\}, \quad a_2(n) = \sum_{k=0}^{n} \alpha(k), \quad Q_1(n) = \alpha(n) a_1(n),$$

$$Q_2(n) = P\{\mu(n) > 0\} a_2(n), \quad \theta_1(n) = \frac{Q_1(n)}{Q_2(n)}, \quad \theta_2(n) = Q_1(n) - Q_2(n),$$

$$a_1 = a_1(\infty),$$

$$\pi_1(\theta, x) = \begin{cases} 0 & , x \le 0 \\ \dfrac{x\theta}{1+\theta} & , 0 < x \le 1 \\ 1 & , x > 1 \end{cases}, \quad \pi_2(\theta, x) = \begin{cases} 0 & , x \le 0 \\ \dfrac{\theta + 1 - e^{-x}}{1+\theta} & , x > 0 \end{cases},$$

$$\pi_3(c_1, c_2, x) = \frac{c_2}{c_1(1+c_2)} \sum_{k \le x} G(1, k), \quad \pi_4(c_1, c_2, x) = \begin{cases} 0 & , x \le 0 \\ \dfrac{c_1 + \pi_5(c_2, x)}{1+c_1} & , x > 0 \end{cases}$$

and the distribution $\pi_5(c, x)$ satisfies the equation:

III. Time-Dependent Immigration

$$\int_0^\infty e^{-\tau x} d\pi_5(c,x) = 1-\tau(1+\tau^c)^{1/c}, \quad c\in(0,1].$$

The following function will be used under normalizations:

$$T(x) = \exp\left\{\int_0^x g(u)du\right\}, \quad \Omega(x)=T(u(1-x^{-1})). \tag{1.10}$$

Since $U(s)\to\infty$ as $s\to 1$, it follows that $\Omega(x)\to\infty$ as $x\to\infty$. It follows from (1.3) that $U(1-x^{-1})$ is a regularly varying function as $x\to\infty$. Therefore, since the $\ln T(x)$ is slowly varying as $x\to\infty$ when $a_1=\infty$, the function $\ln\Omega(x)$ is slowly varying at infinity. For example, if $\nu=1$ and $L(x)\to C\in(0,\infty)$ as $x\to 0$, then we obtain from (1.3)-(1.5) that $\Omega(x)\sim c^{-1}x^{1/c}$ as $x\to\infty$. But, if $L(1/x)=\ln x$, $x>0$, then $\Omega(x)\sim\ln x$ as $x\to\infty$.

In this section we show that under some conditions $Z(n)\in \bigcup_{i=1}^{6} A_i$, $n\in N_0$, where

$$A_1=A\left(\left(\frac{\Omega(y)}{\Omega\left(\frac{1}{g(n)}\right)}\right)^{\alpha(n)}, x\right), \quad A_2=A\left(\frac{\Omega^{\alpha(n)}(y)-1}{\Omega^{\alpha(n)}\left(\frac{1}{g(n)}\right)-1}, x\right),$$

$$A_3=A\left(\frac{\ln\Omega(y)}{\ln\Omega\left(\frac{1}{g(n)}\right)}, \pi_1(\theta_1(n),x)\right), \quad A_4=A(yg(n), \pi_2(\theta_1(n),x)),$$

$$A_5=A(y, \pi_3(a_1,\theta_1(n),x)), \quad A_6=A(yg(n), \pi_4(\theta_1(n),\nu,x)).$$

Assume that there exist the following (finite or infinite) limits:

$$\lim_{n\to\infty} a_1(n)=a_1, \quad \lim_{n\to\infty} Q_1(n) = c, \tag{1.11}$$

$$\beta(n)=o(\theta_2(n)), \quad n\to\infty, \tag{1.12}$$

and denote

$$K_1 = \{\theta_1(n): \liminf_{n\to\infty} \theta_1(n)>0\}, \quad K_2=\{\theta_1(n): \limsup_{n\to\infty} \theta_1(n)<\infty\}.$$

Theorem 1.1. Let conditions (1.1), (1.8), (1.11), (1.12) be satisfied.

1°. If there is the $\lim_{n\to\infty}\theta_2(n) = \theta_2\in[0,\infty]$, then

$$\lim_{n\to\infty} P\{z(n)>0\}(1-e^{-\theta_2(n)})^{-1} = 1;$$

§3.1. The Decreasing Immigration

2°. $Z(n) \in A$ and the components of the pair are defined in the following way. Let $a_1 = \infty$. If $c = \infty$, then $A = A_1$; if $c \in (0, \infty)$, then $A = A_2$; if $c = 0$ and $\theta_1 \in K_1$, then $A = A_3$; if $c = 0$, but $\theta_1 \in K_2$, then $A = A_4$. Now let $a_1 < \infty$. If $\theta_1 \in K_1$, then $A = A_5$; if $\theta_1 \in K_2$, then $A = A_6$.

Remark 1.1. It follows from part 1° of Theorem 1.1 that the probability $P\{Z(n) > 0\}$ tends to 1, to $1 - e^{-\theta_2}$ and to zero, if $\theta_2 = \infty$, $\theta_2 \in (0, \infty)$ and $\theta_2 = 0$, respectively. In addition, in the last case the relation $P\{Z(n) > 0\} \sim \theta_2(n)$ is true as $n \to \infty$.

Remark 1.2. If there exists the limit of $\theta_1(n)$ when $n \to \infty$, we obtain the full spectrum of limit theorems for $Z(n)$ from part 2°. It is interesting to note that, if this limit does not exist, but $\theta_1(n) \in K_1 \cap K_2$, then $Z(n) \in A_3 \cap A_4$ when $a_1 = \infty$, $c = 0$ and $Z(n) \in A_5 \cap A_6$, when $a_1 < \infty$. In particular, if $\theta_1(n) \to \theta_1 \in (0, \infty)$, then the process has different limit distributions under different normalizations: $\pi_1(\theta_1, x)$ and $\pi_2(\theta_1, x)$ when $a_1 = \infty$, $c = 0$; $\pi_3(a_1, \theta_1, x)$ and $\pi_4(\theta_1, \nu, x)$, when $a_1 < \infty$.

We now mention some theorems that obviously explain this phenomenon. To do this, we define a "partial process" $Z_n(l, m)$ in the following way. Let $\zeta_n(k)$ be the number of particles at time n generated by particles immigrating at the time k. We put for any n, e, $m \in N_0$, $l \le m$,

$$Z_n(l, m) = \sum_{k=l}^{m} \zeta_n(k). \tag{1.13}$$

Since $\alpha(n)$ is the regularly varying function, it can be represented in the form $\alpha(n) = l(n) n^{-\alpha}$, $n \in N$, $\alpha \ge 0$ where $l(n)$ is a slowly varying function as $n \to \infty$.

Denote by $L_i(n)$, $i = 1, 2$, positive, integer-valued functions such that $L_i \to \infty$, $L_i(n) = o(n)$ as $n \to \infty$ and

$$a_2(L_1(n)) = a_2(n)(1 + o(1)), \quad \alpha \ge 1, \tag{1.14}$$

$$a_1(L_2(n)) = a_1(n)(1 + o(1)), \quad n \to \infty. \tag{1.15}$$

The existence of such functions follows from definitions of $\alpha(n)$ and $1 - f_n(0)$. We put:

$$Z_0(n) = Z_n(0, L_1(n)), \quad Z_1(n) = Z_n(L_1(n) + 1, n - L_2(n) - 1),$$

$$Z_2(n) = Z_n(n - L_2(n), n), \quad F_i(n, s) = Ms^{Z_i(n)}, \quad |s| \le 1, \quad i = 0, 1, 2,$$

$$B_i = \{Z_i(n) > 0, \ Z_j(n) = 0, \ j = 0, 1, 2, \ j \ne i\}, \quad i = 0, 1, 2.$$

Theorem 1.2. Let conditions (1.1) and (1.8) be fulfilled, and let $a_1=\infty$ and $Q_1(n) \to 0$ as $n \to \infty$.

1^O If $\beta(n)=o(Q_1(n))$, then

$$\lim_{n \to \infty} P\left\{\frac{\ln \Omega(Z(n))}{\ln \Omega(1/g(n))} \leq x \mid B_2\right\} = x, \quad 0 \leq x \leq 1;$$

2^O if $\alpha \geq 1$, then

$$\lim_{n \to \infty} P\{Z(n)(1-f_n(0)) \leq x \mid B_o\} = \begin{cases} 0, & x<0, \\ 1-e^{-x}, & x \geq 0. \end{cases}$$

Theorem 1.3. Let conditions (1.1) and (1.8) be fulfilled, and $a_1 < \infty$.

1^O If $\beta(n)=o(Q_1(n))$, then there are the limits:

$$\lim_{n \to \infty} P\{Z(n)=k \mid B_2\} = P_k^*, \quad \sum_{k=1}^{\infty} P_k^* = 1, \qquad (1.16)$$

and the generating function of P_k^* is

$$\varphi(s) = \frac{1}{a_1} \sum_{k=0}^{\infty} (f_k(s) - f_k(0)); \qquad (1.17)$$

2^O if $\alpha \geq 1$, then

$$\lim_{n \to \infty} P\{Z(n)(1-f_n(0)) \leq x \mid B_o\} = \pi_5(\nu,x).$$

We now turn to prove Theorem 1.1. The proof consists of two steps. In the first step we investigate by analytical methods the partial processes introduced above. In the second step we deduce limit theorems for $Z(n)$ from results obtained for the partial processes by direct probability arguments.

This scheme of the proof allows us to find a family of "accompanying" distributions in the case when the continuous theorem for Laplace transforms cannot be used and allows us to obtain "thinner" results explaining the phenomenon described in Remark 1.2. In addition this scheme will be used in order to study local probabilities in §3.3.

First we prove some lemmas on asymptotics of probability $P(n)=P\{Z(n)>0\}$.

Lemma 1.1. a) If $n \to \infty$, $\alpha(n) \to 0$, $a_1 < \infty$, then $P(n) \to 0$.

b) Let $n \to \infty$, $\alpha(n) \to 0$ and $a_1 = \infty$.

1^O If $Q_1(n) \to \infty$, then $P(n) \to 1$;

2^O if $Q_1(n) \to C \in (0,\infty)$, then $P(n) \to 1-e^{-C}$;

3° if $Q_1(n) \to 0$, then $P(n) \to 0$.

§3.1. The Decreasing Immigration

Proof. We start from part a). Using (1.6) we obtain that $F_n(0)=e^{\Delta_n}$, where

$$0 \geq \Delta_n = \sum_{k=0}^{n} \ln h_k(f_{n-k}(0)) \geq -\sum_{k=0}^{n} \frac{1-h_k(f_{n-k}(0))}{h_k(f_{n-k}(0))}. \tag{1.18}$$

It follows from the relation

$$1-h_k(s) \leq \alpha(k)(1-s) \tag{1.19}$$

that

$$\min_{0 \leq k \leq n} h_k(f_{n-k}(0)) \geq 1 - \max_{0 \leq k \leq n} \alpha(k)(1-f_{n-k}(0)) \to 1, \quad n \to \infty. \tag{1.20}$$

In addition, by (1.17) and (1.19) we have:

$$\sum_{k=0}^{n} (1-h_k(f_{n-k}(0))) \leq \sum_{k=0}^{n} \alpha(k)g(n-k). \tag{1.21}$$

Consider the sum on the right-hand side of (1.21). It is not difficult to show that this sum tends to zero. To do this it is sufficient to decompose it in two sums: from zero to $[n/\ln n]$ and from $[n/\ln n]+1$ to n. Then we obtain from relations (1.18), (1.20) and (1.21) that $\Delta_n \to 0$, that is, $P\{Z(n)=0\}=F_n(0) \to 1$ as $n \to \infty$.

Part b) of Lemma 1.1 follows from the following statement.

Lemma 1.2. Let $n \to \infty$, $a_1(n) \to \infty$, $\nu=1$ and c be a positive constant. If $Q_1(n) \geq c$ for $n \geq N_0$, where N_0 is some positive integer, then

$$\liminf_{n \to \infty} P\{Z(n)>0\} \geq 1-e^{-c+c_0},$$

where $0<c_0<\infty$ and $c_0=0$, if $\beta(n) \to 0$; but if $Q_1(n) \leq c$ for $n \geq N_0$, then

$$\limsup_{n \to \infty} P\{Z(n)>0\} \leq 1-e^{-c}.$$

The proof of Lemma 1.2 is analogous to the proof of the following lemma and, therefore, we omit it.

Lemma 1.3. Let $n \to \infty$, $\alpha(n)=l(n)/n^{\alpha}$, $\alpha \geq 0$, $n \geq 1$, where $l(n)$ is a slowly varying function and $l(n)=o(1/a_1(n))$, when $\alpha=0$.

1° if $\beta(n)=o(Q_1(n))$, then

$$P_1=P\{Z_n(n-L_2(n),n)>0\} \sim Q_1(n);$$

2° a) if $\alpha \geq 1$ and $\beta(n)=O(1)$, then

$$P_2=P\{Z_n(0,L_1(n))>0\} \sim Q_2(n);$$

b) if $\alpha<1$, then $P_2=o(Q_2(n))$.

Proof. Since the random variables $\zeta_n(k)$, $k \in N_0$, are independent, it follows from (1.13) that

$$MS^{Z_n(1,m)} = \prod_{k=1}^{m} h_k(f_{n-k}(s)). \tag{1.22}$$

By virtue of (1.20) and (1.22) as $n \to \infty$

$$\ln P\{Z_n(n-L_2(n),n)=0\} = -(1+o(1)) \sum_{k=n-L_2(n)}^{n} (1-h_k(f(0))). \tag{1.23}$$

From (1.9) and (1.23) we obtain the equality

$$\ln P\{Z_n(n-L_2(n),n)=0\} = -(I_1+I_2)(1+o(1)), \quad n \to \infty, \tag{1.24}$$

where

$$I_1 = \sum_{k=n-L_2(n)}^{n} \alpha(k)(1-f_{n-k}(0)),$$

$$I_2 = \sum_{k=n-L_2(n)}^{n} 2^{-1}\hat{\beta}_k(f_{n-k}(0))(1-f_{n-k}(0))^2.$$

It follows from the definition of $\alpha(n)$ and relation (1.15) that

$$I_1 \sim Q_1(n), \tag{1.25}$$

and, since $\sum_{k=0}^{\infty} (1-f_k(0))^2 < \infty$ and $\beta(n)=o(Q_1(n))$, then

$$I_2 = o(Q_1(n)). \tag{1.26}$$

The statement of part 1° follows from relation $1-e^{-x} \sim x$, $x \to 0$, and (1.24)-(1.26).

Proof of part 2°. It follows from (1.9), (1.20) and (1.22) that as $n \to \infty$

$$\ln P\{Z_n(0,L_1(n))=0\} = -(\hat{I}_1+\hat{I}_2)(1+o(1)) = O(\hat{I}_1), \tag{1.27}$$

where

$$\hat{I}_1 = \sum_{k=0}^{L_1(n)} \alpha(k)(1-f_{n-k}(0)),$$

$$\hat{I}_2 = \sum_{k=0}^{L_1(n)} 2^{-1}\hat{\beta}_k(f_{n-k}(0))(1-f_{n-k}(0))^2.$$

In the case of $\alpha \geq 1$ we obtain from (1.27) that

$$\ln P\{Z_n(0,L_1(n))=0\} \sim -Q_2(n), \quad n \to \infty. \tag{1.28}$$

If $\alpha<1$, then $\sum_{k=0}^{L_1(n)} \alpha(k)=o(a_2(n))$ as $n \to \infty$ and, consequently,

$$\hat{I}_1 = (1+o(1))(1-f_n(0)) \sum_{k=0}^{L_1(n)} \alpha(k) = o(Q_2(n)),$$

§3.1. The Decreasing Immigration

and by virtue of (1.27) we get:

$$\ln P\{Z_n(0,L_1(n))=0\} = o(Q_2(n)), \quad n\to\infty. \tag{1.29}$$

The statement of part 2° follows from (1.28) and (1.29). The lemma is proved.

Lemma 1.4. Let $S_n = e^{-\tau c_n}$, $\tau>0$, $c_n \to 0$ as $n\to\infty$, $U(1-c_n)=O(n)$, and let $L_2(n)$ be a positive integer-valued function such that $L_2(n)\to\infty$, $L_2(n)=O(n)$ as $n\to\infty$ and satisfies the condition (1.15). Then

$$\sum_{k=0}^{L_2(n)-1} g(k+U(S_n)) = (c_1(n)+c_2(n))(1+o(1)), \quad n\to\infty, \tag{1.30}$$

where

$$c_1(n) = \ln\frac{\Omega(1/g(n))}{\Omega(1/c_n)} + O(1), \quad n\to\infty, \tag{1.31}$$

$$0 \le c_2(n) \le g(U(S_n)) - g(n+U(S_n)). \tag{1.32}$$

Proof. We consider the following relation

$$\sum_{k=0}^{n-1} g(k+U(S_n)) = c_1(n)+c_2(n), \tag{1.33}$$

where

$$c_1(n) = \int_0^n g(x+U(S_n))dx,$$

$$c_2(n) = \sum_{k=0}^{n-1} [g(k+U(S_n)) - \int_k^{k+1} g(x+U(S_n))dx].$$

Inequality (1.32) follows from the definition of $c_2(n)$ and the monotonicity of $g(x)$. In order to prove (1.31), note that

$$c_1(n) = \ln\frac{\Omega(1/g(n))}{\Omega(1/c_n)} + \ln\frac{T(U(S_n)+n)}{T(n)} + \ln\frac{T(U(1-c_n))}{T(U(S_n))}. \tag{1.34}$$

We show that the last two summands in (1.34) are $O(1)$ as $n\to\infty$. In fact, using Lagrange's formula, we have:

$$0 \le \ln\frac{T(U(S_n)+n)}{T(n)} \le \exp\left\{\int_n^{n+U(S_n)} g(x)dx\right\} - 1 < e^{g(n)U(S_n)}. \tag{1.35}$$

By virtue of (1.3) as $n\to\infty$

$$U(S_n) \sim \tau^{-1}U(1-c_n)=O(n). \tag{1.36}$$

Granting relations (1.4), (1.5) and (1.36), we obtain from (1.35) that the first term in (1.34) is $O(1)$ as $n \to \infty$.

Again using Lagrange's formula, we have:

$$\ln \frac{T(U(1-c_n))}{T(U(S_n))} = \frac{T(U(1-c_n))-T(U(S_n))}{T(U(S_n))+[T(U(1-c_n))-T(U(S_n))]x_o}, \qquad (1.37)$$

where $n \in N$, $0 < x_o < 1$. Consider the two cases. If $1-c_n \geq U(S_n)$, then we obtain from (1.37) that

$$0 \leq \ln \frac{T(U(1-c_n))}{T(U(S_n))} < \exp\{g(U(S_n))(U(1-c_n)-U(S_n))\}.$$

Granting relations (1.4), (1.5) and (1.36), we obtain that the expression under the exponential curve is $O(1)$ as $n \to \infty$. But, if $1-c_n < U(S_n)$, then using (1.37) we have:

$$0 \geq \ln \frac{T(U(1-c_n))}{T(U(S_n))} \geq 1-\exp\{g(U(1-c_n))[U(S_n)-U(1-c_n)]\},$$

where the expression under the exponential curve is again $O(1)$.

In order to complete the proof, it remains to show that

$$\sum_{k=0}^{L_2(n)-1} g(k+U(S_n)) = (1+o(1)) \sum_{k=0}^{n-1} g(k+U(S_n)), \quad n \to \infty.$$

But this relation follows from the choice of function $L_2(n)$ and from the fact that $\ln T(x)$ is a slowly varying function as $x \to \infty$. Lemma 1.4 is proved.

Lemma 1.5. If $n \to \infty$, $S_n \to 1$, $(1-S_n) \max\limits_{0 \leq k \leq n} \alpha(k) \to 0$ and $\sum\limits_{k=0}^{n} (1-h_k(f_{n-k}(S_n)))=O(1)$, $n \to \infty$, then

$$F_n(S_n) \sim \exp\left\{-\sum_{k=0}^{n} (1-h_k(f_{n-k}(S_n)))\right\}.$$

The statement of this lemma follows from relations (1.6) and (1.20).

We denote by $T^{-1}(\cdot)$ the inverse function of $T(x)$ and put $0 < x < 1$,

$$K(n)=T^{-1}([x(T^{\alpha(n)}(n)-1)+1]^{1/\alpha(n)}).$$

It is clear that $K(n) \to \infty$ as $n \to \infty$.

Lemma 1.6. If $a_1=\infty$, $n \to \infty$, $\alpha(n) \to 0$ and is slowly varying, and $\beta(n) \to 0$, then for any $\tau > 0$

§3.1. The Decreasing Immigration

$$M \exp\{-\tau g(K(n))Z(n)\} \sim [T(K(n))/T(n)]^{\alpha(n)}.$$

Proof. Let $\lambda_\alpha(n)$ be a function such that $\lambda_\alpha(n) \to \infty$, $\alpha(n/\lambda_\alpha(n)) \sim \alpha(n)$ as $n \to \infty$ (see Badalbaev and Rahimov (1978)). If we put $S_n = \exp\{-\tau g(K(n))\}$, $\tau > 0$, then

$$A_1(n) = \sum_{k=[n/\lambda_\alpha(n)]}^{n} \alpha(k) g(n-k+U(S_n))$$

$$= (1+o(1))\alpha(n) \sum_{k=0}^{n-[n/\lambda_\alpha(n)]} g(k+U(S_n)), \quad n \to \infty,$$

and, using Lemma 1.4, we obtain:

$$A_1(n) = \alpha(n)[c_1(n) + c_2(n)](1+o(1)), \quad n \to \infty,$$

where $c_2(n) \to 0$ and

$$c_1(n) = \frac{T(n)}{T(K(n))} + O(1).$$

By virtue of (1.9) we have:

$$A_2(n) \sum_{k=0}^{n} [1 - h_k(f_{n-k}(S_n))] = A_1(n)$$

$$+ \sum_{k=0}^{[n/\lambda_\alpha(n)]} \alpha(k) g(n-k+U(S_n)) + O(\sum_{k=0}^{n} \beta(k)[g(n-k+U(S_n))]^2) \quad (1.38)$$

Since $\alpha(n) \to 0$, $\beta(n) \to 0$ as $n \to \infty$, the last two terms in (1.38) tend to zero and, therefore,

$$e^{-A_2(n)} \sim e^{-A_1(n)}. \quad (1.39)$$

The statement of the lemma follows from relations (1.38), (1.39) and Lemma 1.5.

Proof of Theorem 1.2. It is obvious that

$$M\left[S^{Z(n)} | B_i\right] = M\left[S^{Z_i(n)} | Z_i(n) > 0\right] = 1 - \frac{1 - F_i(n,s)}{P\{Z_i(n) > 0\}}, \quad i = 0, 1, 2. \quad (1.40)$$

Consider (1.40) with $i=2$, $S_n = \exp\{-\tau g(K(n))\}$, $0 < x < 1$, $K(n) = T^{-1}(T^x(n))$. By virtue of (1.22) as $n \to \infty$

$$\ln F_2(n, S_n) = -(1+o(1)) \sum_{k=n-L_2(n)}^{n} (1 - h_k(f_{n-k}(S_n))) \quad (1.41)$$

and, using (1.7) and (1.9), we obtain:

$$\ln F_2(n, S_n) = -(I_1 + I_2)(1+o(1)), \quad (1.42)$$

where

III. Time-Dependent Immigration

$$I_1 = \sum_{k=n-L_2(n)}^{n} \alpha(k)g(n-k+U(S_n)),$$

$$I_2 = \sum_{k=n-L_2(n)}^{n} \hat{\beta}_k(f_{n-k}(S_n))2^{-1}[g(n-k+U(S_n))]^2.$$

It follows from definition $\alpha(n)$ that as $n \to \infty$

$$I = (1+o(1))\alpha(n) = \sum_{k=0}^{L_2(n)} g(k+U(S_n)). \qquad (1.43)$$

Since under the conditions of part 1° the function $T(U(1-g(K(n))))= T(K(n))= =T^x(n)$ is $o(T(n))$ as $n \to \infty$, we can apply Lemma 1.4 to the sum in (1.43). Therefore,

$$I_1 = \alpha(n)(c_1(n)+c_2(n)(1+o(1))),$$

where

$$c_1(n) \sim \ln \frac{\Omega(1/(1-f_n(0)))}{\Omega(1/g(K(n)))} = (1-x)\ln T(n),$$

$$c_2(n) \leq \tau g(K(n)) \leq (\tau/K(n))\ln T(K(n)) = (\tau x/K(n))\ln T(n),$$

and, since $K(n) \to \infty$, we have:

$$I_1 = (1-x)Q_1(n)(1+o(1)). \qquad (1.44)$$

Since $\sum_{k=0}^{\infty} (1-f_k(0))^2 < \infty$ and $\beta(n)=o(Q_1(n))$ under the conditions of part 1°,

$$I_2 = o(Q_1(n)), \quad n \to \infty. \qquad (1.45)$$

If we substitute (1.41) and (1.44) into (1.42), granting that $Q_1(n) \to 0$ as $n \to \infty$ we obtain:

$$1-F_2(n,S_n)=(1-\sigma)Q_1(n)(1+o(1)). \qquad (1.46)$$

It follows from relations (1.40), (1.46), part 1° of Lemma 1.3 and the continuity theorem for Laplace transforms that

$$P\{Z(n)g(K(n)) \leq 1 \mid B_2\} \to x, \quad n \to \infty.$$

Thus part 1° is proved.

Now let us prove part 2°. To do this, consider $F_o(n,S_n)$ with $S_n = \exp\{-\tau(1-f_n(0))\}$, $\tau>0$. Using relations (1.7), (1.9), (1.30), we obtain that

$$\ln F_o(n,S_n) = -(I_1+I_2)(1+o(1)), \quad n \to \infty,$$

where

§3.1. The Decreasing Immigration

$$I_1 = \sum_{k=0}^{L_1(n)} \alpha(k)g(n-k+U(S_n)),$$

$$I_2 = \sum_{k=0}^{L_1(n)} \hat{\beta}_k(f_{n-k}(S_n))[g(n-k+U(S_n))]^2.$$

Since $g(x)=x^{-1}N(x)$, granting the choice of the function $L_1(n)$, we have

$$I_1 = g(n+U(S_n))(1+o(1))\sum_{k=0}^{n}\alpha(k), \quad n\to\infty.$$

When $S_n = e^{-\tau g(n)}$, it follows from (1.3) that

$$U(S_n) \sim n\tau^{-\nu} \qquad (1.47)$$

and, consequently, $I_1 = Q_2(n)(\tau/(1+\tau))(1+o(1))$. Since $L_1(n)=o(n)$ and $1-f_n(o)=O(1/n)$ the following relation

$$I_2 = O((1-f_n(0))^2 L_1(n)) = o(Q_2(n)), \quad n\to\infty,$$

is true. Thus,

$$1-F_o(n,S_n) = Q_2(n)(\tau/(1+\tau))(1+o(1)), \quad n\to\infty.$$

If we substitute the last relation into (1.40), using part 2 a) of Lemma 1.3 we obtain:

$$M\left[S_n^{Z(n)}|B_o\right] \to (1+\tau)^{-1}$$

as $n\to\infty$ and for $\tau>0$, $S_n = e^{-\tau g(n)}$.

Theorem 1.2 is proved.

Since the proof of Theorem 1.3 is just like that of Theorem 1.2, we will represent a scheme of this proof without the technical details.

Proof of Theorem 1.3. Let us prove part 1°. Consider relation (1.40) with $i=2$. Using (1.7), (1.9) and granting lemmas 1.1-1.3, we obtain that as $n\to\infty$ and $0\leq s\leq 1$

$$\ln F_2(n,s) = -(I_1+I_2)(1+o(1)), \qquad (1.48)$$

where

$$I_1 = \sum_{k=n-L_2(n)}^{n} \alpha(k)g(n-k+U(s)),$$

$$I_2 = \sum_{k=n-L_2(n)}^{n} 2^{-1}\hat{\beta}_k(f_{n-k}(s))[g(n-k+U(s))]^2.$$

It is obvious that

III. Time-Dependent Immigration

$$I_1 = (1+o(1)) \sum_{k=0}^{\infty} g(k+U(s)), \quad n \to \infty. \tag{1.49}$$

Using the corresponding estimates of I_2 from the proof of Lemma 1.3, by virtue of the monotonicity of the function $g(x)$, we obtain:

$$I_2 = o(Q_1(n)), \quad n \to \infty. \tag{1.50}$$

Substituting (1.48)-(1.50) into relation (1.40) and granting the result of part 1° of Lemma 1.3, we obtain the relation

$$M[S^{Z_1(n)} \mid z_1(n) > 0] \to \varphi(s), \quad n \to \infty,$$

thus the statement of part 1° holds.

Proof of part 2°. Consider (1.40) with $i=0$, $S = S_n = e^{-\tau g(n)}$, $\tau > 0$. As in the proof of part 1°, we have:

$$\ln F_o(n, S_n) = -(I_1 + I_2)(1+o(1)), \tag{1.51}$$

where

$$I_1 = \sum_{k=0}^{L_1(n)} \alpha(k) g(n-k+U(S_n)),$$

$$I_2 = \sum_{k=0}^{L_1(n)} 2^{-1} \hat{\beta}_k (f_{n-k}(S_n)) [g(n-k+U(S_n))]^2.$$

Taking into account the choice of the function $L_1(n)$, we obtain from relations (1.4) and (1.47) that as $n \to \infty$

$$I_1 = Q_2(n) \frac{\tau(1+o(1))}{(1+\tau^\nu)^{1/\nu}}, \quad I_2 = o(Q_2(n)). \tag{1.52}$$

We conclude from (1.40), (1.51) and (1.52) that $M[S^{Z_o(n)}_n \mid Z_o(n) > 0]$ tends to the Laplace transform of the distribution $\pi_5(\nu, x)$. Theorem 1.3 is proved.

§1b. The Proof of the Main Theorem

Proof of Theorem 1.1. Proof of part 1°. When $\theta_2 = \infty$ and $\theta_2 \in (0, \infty)$ the statement of this part follows from Lemma 1.1. Therefore it is sufficient to show that

$$P(n) \sim \theta_2(n), \quad n \to \infty, \tag{1.53}$$

when $\theta_2(n)$ tends to zero. It is obvious that

$$\{Z(n) > 0\} = \{Z_o(n) > 0\} \cup \{Z_1(n) > 0\} \cup \{Z_2(n) > 0\}. \tag{1.54}$$

§3.1. The Decreasing Immigration

Consider the case $1-h_k(s)=\alpha(k)(1-s)$. Put

$$P_o \equiv P\{Z_n(L_1(n)+1, n-L_2(n)-1)>0\} = 1 - \prod_{k=L_1(n)+1}^{n-L_2(n)-1} h_k(f_{n-k}(0)). \tag{1.55}$$

Since

$$\prod_i (1-\varepsilon_i) \geq 1 - \sum_i \varepsilon_i, \quad i \geq 0,$$

the following estimation is true

$$P_o \leq \sum_{k=L_1(n)+1}^{n-L_2(n)-1} \alpha(k)(1-f_{n-k}(0)). \tag{1.56}$$

It follows from assumptions about $f_n(s)$ and $\alpha(k)$ that for a sufficiently large n

$$\max_{0 \leq k \leq n} \left\{ \frac{1-f_n(0)}{1-f_{n-k}(0)}, \frac{\alpha(n)}{\alpha(k)} \right\} \geq c(\alpha) > 0,$$

and, therefore,

$$P_o = O\left(\alpha(n) \sum_{k=L_1(n)}^{n-L_2(n)} (1-f_{n-k}(0)) + (1-f_n(0)) \sum_{k=L_1(n)+1}^{n-L_2(n)-1} \alpha(k) \right). \tag{1.57}$$

The first term of (1.57) is not greater than

$$\alpha(n) \sum_{k=L_1(n)}^{n-L_2(n)} (1-f_{n-k}(0)) = o(Q_1(n)), \tag{1.58}$$

and the second term is less than

$$(1-f_n(0)) \sum_{k=L_1(n)}^{n-L_2(n)} \alpha(k) = \begin{cases} o(Q_2(n)), & \text{if } \alpha \geq 1, \\ O(Q_2(n)) = o(Q_1(n)), & \text{if } \alpha < 1 \end{cases}. \tag{1.59}$$

We obtain from (1.57)-(1.59) that as $n \to \infty$

$$P_o = o(Q_1(n) + Q_2(n)). \tag{1.60}$$

Since the events on the right-hand side of (1.54) are independent

$$P\{(Z(n)>0\} = \sum_{i=0}^{2} P_i - \sum_{i \neq j} P_i P_j + P_o P_1 P_2. \tag{1.61}$$

In order to obtain (1.53) in the case $1-h_k(s)=\alpha(k)(1-s)$, it is sufficient to substitute asymptotic representations of P_o, P_1 and P_2 from (1.60) and Lemma 1.3 into (1.61). In the general case we use equality (1.9) and the relation

III. Time-Dependent Immigration

$$\sum_{k=0}^{n} \beta(k)(1-f_{n-k}(0))^2 = o(Q_1(n)+Q_2(n)),$$

which is true when $\beta(n)=o(Q_2(n))$. Part 1° of Theorem 1.1 is proved.

Proof of part 2. Let $a_1=\infty$, $c=\infty$. Then $T^{\alpha(n)}(n) \to \infty$ and $[T(K(n))/T(n)]^{\alpha(n)} \to x$, $0<x<1$, where $K(n)$ is the same as in Lemma 1.6. We obtain from Lemma 1.6 that

$$\lim_{n \to \infty} M \exp\{-\tau g(K(n))Z(n)\} = x, \quad \tau>0, \tag{1.62}$$

and, consequently,

$$\lim_{n \to \infty} P\{Z(n)g(K(n)) \leq 1\} = x. \tag{1.63}$$

Since

$$P\{Z(n)g(K(n)) \leq 1\} = P\left\{\left[\frac{\Omega(Z(n))}{T(n)}\right]^{\alpha(n)} \leq x+(1-x)T^{-\alpha(n)}(n)\right\},$$

$$T(n) = T(U(1-g(n))) = \Omega(1/(1-f_n(0))),$$

we obtain from (1.63) that $Z(n) \in B([\Omega(y)/T(n)]^{\alpha(n)}, x)$. Hence it follows from Lemma 1.1 that $Z(n) \in A$.

Let now $a_1=\infty$, $c \in (0,\infty)$. We have to prove that

$$\lim_{n \to \infty} P\left\{\frac{[\Omega(Z(n))]^{\alpha(n)}-1}{[\Omega(1/(1-f_n(0)))]^{\alpha(n)}-1} \leq x \mid Z(n)>0\right\} = x, \quad 0 \leq x \leq 1. \tag{1.64}$$

In this case $T^{\alpha(n)}(n) \to e^c$, $\theta_2(n) \sim Q_1(n)$, $n \to \infty$, and

$$[T(K(n))/T(n)]^{\alpha(n)} \to x(1-e^{-c})+e^{-c}, \quad n \to \infty.$$

Considering the equality

$$M[\exp\{-\tau g(K(n))Z(n)\} \mid Z(n)>0] = \frac{M[\exp\{-\tau g(K(n))Z(n)\}] - P\{Z(n)=0\}}{1-P\{Z(n)=0\}}$$

and using Lemmas 1.1 and 1.6, we have

$$M[\exp\{-\tau g(K(n))Z(n)\} \mid Z(n)>0] \to x, \quad n \to \infty. \tag{1.65}$$

Relation (1.64) follows from (1.65). But relation (1.64) is equivalent to the statement $Z(n) \in A_2$.

Let $a_1=\infty$, $c=0$, $\theta_1 \in K_1$. Introduce the following events:

$$B_{ij} = \{Z_i(n)>0, Z_j(n)>0, Z_l(n)=0, l=0,1,2; l \neq i, l \neq j\},$$

$$B_{012} = \{Z_0(n)>0, Z_1(n)>0, Z_2(n)>0\}, \quad i \neq j.$$

§3.1. The Decreasing Immigration

It is obvious that

$$\{Z(n)>0\} = \bigcup_{i=0}^{2} B_i \cup \bigcup_{0 \leq i \leq j \leq 2} B_{ij} \cup B_{012}.$$

Then for any event A

$$P\{A|Z(n)>0\} = \sum_{i=0}^{2} P\{A|B_i\} \frac{P\{B_i\}}{P\{Z(n)>0\}} + \sum_{0 \leq i < j \leq 2} P\{A|B_{ij}\} \frac{P\{B_{ij}\}}{P\{Z(n)>0\}}$$

$$+ P\{A|B_{012}\} \frac{P\{B_{012}\}}{P\{Z(n)>0\}}. \qquad (1.66)$$

By virtue of the independence of processes $Z_i(n)$, $i=0,1,2$,

$$P\{B_i\} = P\{Z_i(n)>0\} \prod_{\substack{j=0 \\ j \neq i}}^{2} P\{Z_j(n)=0\},$$

$$P\{B_{ij}\} = P\{Z_i(n)>0\} P\{Z_j(n)>0\} P\{Z_l(n)=0, l \neq i, l \neq j\},$$

$$P\{B_{012}\} = \prod_{i=0}^{2} P\{Z_i(n)>0\}. \qquad (1.67)$$

It follows from (1.67) and Lemma 1.3 that as $n \to \infty$

$$P\{B_0\} \sim Q_2(n), \quad \alpha \geq 1, \quad P\{B_2\} \sim Q_1(n), \qquad (1.68)$$

and the remaining probabilities are $o(Q_1(n)+Q_2(n))$. Let

$$A = \{\ln\Omega(Z(n))/\ln\Omega(1/(1-f_n(0))) < x\}, \quad 0 \leq x \leq 1.$$

Granting equality $Z(n) = Z_0(n) + Z_1(n) + Z_2(n)$ and the independence of $Z_i(n)$, we obtain that

$$P\{A/B_i\} = \left\{ \frac{\ln[\Omega(Z_i(n))]}{\ln\Omega[1/(1-f_n(0))]} \leq x \mid Z_i(n)>0 \right\}, \quad i=0,1,2.$$

If we use the results of Theorem 1.2, we have that as $n \to \infty$

$$P\{A/B_2\} \to x, \quad P\{A/B_0\} \to 0. \qquad (1.69)$$

Substituting asymptotic representations from relations (1.68), (1.69) for quantities on the right-hand side of (1.66) and using part 1° of Theorem 1.1, we can see that

$$P\{A|Z(n)>0\}=x\theta_1(n))(1+o(1))+o(1), \ n\to\infty.$$

Since $\liminf_{n\to\infty} \theta_1(n)>0$, it follows from the last relation that $Z(n)\in A_3$.

Now let $a_1=\infty$, $c=0$, $\theta_1(n)\in K_2$. We put $A=\{Z(n)(1-f_n(0))\leq x\}$, $0\leq x<\infty$. Then

$$P\{A|B_i\}=P\{(1-f_n(0))Z_i(n)\leq x|Z_i(n)>0\}, \ i=0,1,2.$$

Taking this into account, we obtain from Theorem 1.2 that

$$P\{A/B_2\}\to 1, \ P\{A/B_0\}\to 1-e^{-x}, \ x>0, \ n\to\infty. \tag{1.70}$$

Since $\limsup_{n\to\infty} \theta_1(n)<\infty$, $\alpha\geq 1$. Substituting (1.68) and (1.70) into (1.66), we have as $n\to\infty$

$$P\{A|Z(n)>0\} = \frac{\theta_1(n)}{1+\theta_1(n)}(1+o(1)) + \frac{1-e^{-x}}{1+\theta_1(n)} + o(1),$$

and we can see from this relation that $Z(n)\in A_4$.

In order to show the theorem in the case $a_1<\infty$, it is sufficient to put in relation (1.66) $A=\{Z(n)=k\}$, when $\theta_1(n)\in K_1$ and $A=\{(1-f_n(0))Z(n)\leq x\}$, when $\theta_1(n)\in K_2$. We shall use Theorem 1.3, Lemma 1.3 and part 1° of Theorem 1.1 for an estimation of quantities in (1.66). Theorem 1.1 is proved.

§1c. State-Dependent Immigration

The results of the present section have been proved by Rahimov (1986). The branching process with decreasing immigration was first considered by Badalbaev and Rahimov (1978, 1983) in the case of finite variance. After this many papers studying branching processes with state-dependent decreasing immigration or decreasing migration were published (Mitov and Yanev 1984, 1985; Yanev and Mitov 1982, 1985). In papers by Rahimov and Kaverin (1986) and by Rahimov and Kurbanov (1989) some new results were also obtained for these models. Our final remarks are about such processes.

Let $\mu(n)$ be the critical Galton-Watson process considered above. Define the process $W(n)$ by the following relation

$$W(0)=0, \ W(n+1) = \sum_{i=1}^{W(n)} X_{in} + Y_n\chi(W(n)=0), \tag{1.71}$$

where $\{X_{in}, \ i\in N, \ n\in N_0\}$ are independent and identically distributed random variables having the same distribution as $\mu(1)$, let $\chi(A)$ be the indicator

§3.1. The Decreasing Immigration

function of the event A, as before, and $\sum_{i=1}^{0}(\ )=0$. If we interpret X_{in} and Y_n as the number of direct offspring of the ith particle at time n and as the number of immigrants at time n respectively, then W(n) will be a Galton-Watson process with state-dependent immigration. Models of branching processes where $\{Y_n, n \in N_0\}$ have the same distribution were considered by Foster (1971), Pakes (1971b), Sato (1975) and Yamazato (1975). Limit theorems for W(n) in the case of $f''(1)<\infty$, $MY_n \to 0$ when $n \to \infty$ and $MY_n \ln n \to C$, $0 \le C < 2^{-1} M\mu(1)(\mu(1)-1)$ have been proved by Mitov and Yanev (1984, 1984). In these papers, as in Foster's paper, proofs of theorems are based on analyses of a functional equation for the generating function of W(n) following from (1.71). A lemma from the paper of Rahimov and Kaverin (1986) allows us to represent the following statement.

Proposition 1.1. If $a_1 < \infty$ or $a_1 = \infty$ but $c=0$, then Theorem 1.1 is true with W(n) instead of Z(n).

Now we consider the branching process with non-homogeneous migration. Let two independent families $\{X_{jn}(k),\ k,\ n \in N_0\}$, $j=1,\ 2$, of independent random variables taking values from N_0 be given.

Define the process Y_n, $Y_0=0$, by the relation:

$$Y_n = \begin{cases} \sum_{k=1}^{\max(Y_{n-1}-1,0)} X_{1n-1}(k), & \text{with probability } p_n, \\ \sum_{k=1}^{Y_{n-1}} X_{1n-1}(k) + X_{2n-1}(1), & \text{with probability } r_n, \\ \sum_{k=1}^{Y_{n-1}} X_{1n-1}(k), & \text{with probability } q_n, \end{cases}$$

where $p_n + r_n + q_n = 1$.

Introduce the generating functions:

$$f(S) = MS^{X_{1n}(k)},\quad G(S) = MS^{X_{2n}(1)},\quad H_n(S) = MS^{Y_n},\quad 0 \le S \le 1.$$

Branching processes with migration were studied by Nagayev and Khan (1980) under an assumption of the time homogeneity of the migration. S. Kaverin (1991) got some generalizations of results from the paper of Nagaev and Khan. The process Y_n was considered by Mitov and Yanev (1982, 1985), where limit theorems were proved in the case of $f''(1)<\infty$, $G''(1)<\infty$, when P_n, $r_n \to 0$.

Let $Z(n)$, $Z(0)=0$, be a simple branching process with immigration which has the generating function of the number of direct offspring $f(S)$ and the generating function of the number of immigrants

$$h_n(S)=1-(1-S)P_k H_k(0).$$

We put

$$F_n(S)=MS^{Z(n)}, \quad F_n^*(S)=M[S^{Z(n)}|Z(n)>0], \quad H_n^*(S)=M[S^{Y_n}|Y_n>0],$$

and assume that

$$\left.\begin{array}{l} G(0)>0, \ 0<m=G'(1)<\infty \\ f'(1)=1, \ mr_n \ P_n \to 0, \ n\to\infty \end{array}\right\}. \qquad (1.72)$$

Proposition 1.2. [Rahimov, Kurbanov (1989)]. Let conditions (1.72) be satisfied, $F_n(0)\to 0$ and $1-H_n(0)\sim 1-F_n(0)$ as $n\to\infty$. Then

$$\lim_{n\to\infty} H_n^*(e^{-\lambda K(n)}) = \psi(\lambda) < 1$$

for any $\lambda>0$ and for a positive function $K(n)\to 0$, $n\to\infty$, if and only if

$$\lim_{n\to\infty} F_n^*(e^{-\lambda K(n)}) = \psi(\lambda).$$

Proposition 1.2 allows us to deduce from Theorem 1.1 limit theorems for Y_n in the case where both the number of direct offspring and number of immigrating particles have infinite variances.

§3.2. INCREASING IMMIGRATION

§2a. The Process with Infinite Variance

In this section we consider the case when the expectation $\alpha(n)$ is increasing as $n\to\infty$. Let conditions (1.1) and (1.8) be fulfilled. Define distribution functions $\pi_6(\theta,\nu,x)$ and $\pi_7(c,\nu,x)$ by their Laplace transforms:

$$\int_0^\infty e^{-\tau x} d\pi_6(\theta,\nu,x) = \exp\left\{-\frac{\theta\nu}{1-\nu}\tau^{1-\nu}\right\},$$

$$\int_0^\infty e^{-\tau x} d\pi_7(c,\nu,x) = \exp\left\{-c\int_0^1 \frac{x^{1/\nu-1}}{(1-x-\tau^{-\nu})^{1/\nu}}dx\right\}.$$

Let $g(x)$, $T(x)$, $\Omega(x)$ be functions introduced in §3.1, and let $B(\varphi_t(y), \pi(t,x))$ be the class of processes described by Definition 0.1. Denote by $T^{-1}(\cdot)$ the inverse of the function $T(x)$ and put

§3.2. Increasing Immigration

$$K(n)=T^{-1}(T(n)x^{1/\alpha(n)}), \quad 0<x<1, \quad C_n = \sum_{k=0}^{n} \beta(k).$$

It is easy to see that $K(n)\to\infty$ as $n\to\infty$ when $\alpha(n)\to\infty$.

Assume that there exists the following limit:

$$\lim_{n\to\infty} ng(n)\alpha(n)=c\in[0,\infty), \tag{2.1}$$

$$g(K(n))\alpha(n)=o(1), \quad g^2(K(n))C_n=o(1), \quad n\to\infty. \tag{2.2}$$

Denote by $\psi(x)$, $x\geq 0$, a positive function such that

$$\lim_{n\to\infty} \alpha(n)\psi(n)g(\psi(n))=\theta\in(0,\infty), \quad g^2(\psi(n))C_n=o(1). \tag{2.3}$$

We also put

$$B_1=B\left(\left[\frac{\Omega(y)}{\Omega(1/g(n))}\right]^{\alpha(n)}, x\right), \quad B_2=B(yg(\psi(n)), \pi_6(\theta,\nu,x)),$$

$$B_3=B(yg(n), \pi_7(c,\nu,x)).$$

Theorem 2.1. Let conditions (1.1), (1.8), (2.1) be satisfied and $\alpha(n)\to\infty$. If $c=0$, $\nu=1$ and conditions (2.2) are fulfilled, then $Z(n)\in B_1$; if $c=0$, $\nu<1$ and conditions (2.3) are fulfilled, then $Z(n)\in B_2$; but if $c\in(0,\infty)$ and $g^2(n)C_n\to 0$, then $Z(n)\in B_3$.

Remark 2.1. It is clear that $\pi_6(\theta,\nu,x)$ is a positive stable distribution with parameter $1-\nu$ (see Feller (1967), p.514) and the distribution $\pi_7(c,\nu,x)$, in general, is infinitely divisible.

It is interesting to consider some particular forms of the distribution $\pi_7(c,\nu,x)$. If $\nu=1$, then $\pi_7(c,1,x)$ is a gamma - distribution with density $e^{-x}x^{c-1}/\Gamma(c)$. If $\nu=1/2$, then the limit Laplace transform has the form $(1+\sqrt{\tau})^{-c}\exp\{-c\sqrt{\tau}\}$. In general, if $k\in N$, then $\pi_7(c,(2k)^{-1}x)$ is defined by its Laplace transform:

$$\exp\left\{\frac{c}{2k(1+\tau^{-1})^{2k}} {}_2F_1\left(2k, 2k, 2k+1, \frac{\tau}{1+\tau}\right)\right\},$$

where ${}_2F_1(a,b,c,y)$ is Gauss's hypergeometric function.

Remark 2.2 We shall note that when $c=0$ and $\nu=1$ for $Z(n)\in B_1$ the condition $\alpha(n)\to\infty$ is not necessary, in general. In fact it is sufficient that $K(n)$ tends to infinity as $n\to\infty$. The last condition can be fulfilled, for example, when $\Sigma P(\mu(k)>0=\infty$ and $\alpha(n)\to 0$ not very "fast" (see §2.1). This remark shows that the results of theorems 1.1 and 2.1 complement each other naturally.

III. Time-Dependent Immigration

Proof of Theorem 2.1. Let $c=0$, $\nu=1$ and conditions (2.2) be fulfilled. We consider:

$$I = \sum_{k=0}^{n} (1-h_k(f_{n-k}(S_n))), \quad S_n = e^{-g(K(n))}.$$

It follows from (2.1) and (1.4) that $\alpha(n)$ is a slowly varying function when $\nu=1$. Let $\lambda_\alpha(x) \to \infty$ be a function such that $\alpha(n) \sim \alpha(n/\lambda_\alpha(n))$, $n \to \infty$. Then in the relation

$$I = \sum_{k=0}^{[n/\lambda_\alpha(n)]-1} (1-h_k(f_{n-k}(S_n))) +$$

$$+ \sum_{k=[n/\lambda_\alpha(n)]}^{n} (1-h_k(f_{n-k}(S_n))) = I_1 + I_2 \qquad (2.4)$$

the first term is not greater than

$$\sum_{k=0}^{[n/\lambda_\alpha]} g(n-k+U(S_n))$$

which tends to zero as $n \to \infty$.

Consider I_2. Using relations (1.7) and (1.9), we have:

$$I_2 = \sum_{k=[n/\lambda_\alpha(n)]}^{n} \alpha(k) g(n-k+U(S_n)) +$$

$$+ \frac{1}{2} \sum_{k=[n/\lambda_\alpha(n)]}^{n} \hat{\beta}_k(f_{n-k}(S_n)) g^2(n-k+U(S_n)) = R_1 + R_2. \qquad (2.5)$$

By virtue of the choice of the function $\lambda_\alpha(n)$ and the monotonicity of $g(x)$ we obtain that

$$R_1 = (1+o(1))\alpha(n)[\hat{R}_1(n) + \Delta(n)], \quad n \to \infty, \qquad (2.6)$$

where

$$\hat{R}_1(n) = \int_0^{n-n/\lambda_\alpha(n)} g(x+U(S_n))dx = \ln \frac{T(n-n/\lambda_\alpha(n)+U(S_n))}{T(U(S_n))},$$

$$0 \le \Delta(n) \le g(U(S_n)).$$

Since $U(S_n) \ge U(1-g(K(n))) = K(n)$, we obtain from condition $\alpha(n)g(K(n)) \to 0$ that $\Delta(n)\alpha(n)$ tends to zero as $n \to \infty$.

Using the monotonicity of the function $T(x)$, we have:

§3.2. Increasing Immigration

$$0 \leq T(U(S_n)) - T(K(n)) \leq T'(\theta_n) U'(\eta_n)(g(K(n)) + S_n - 1), \qquad (2.7)$$

where $\theta_n \in [K(n), U(S_n)]$, $\eta_n \in [1-g(K(n)), S_n]$. It follows from the definition of $T(x)$ and the relation

$$U(S) = [\nu(1-s)^\nu M(1-s)]^{-1}, \qquad (2.8)$$

where $M(s) \sim L(s)$, $s \searrow 0$, that $T'(x) = T(x)g(x)$ and $U'(1-x^{-1}) \sim xU(1-x^{-1})$, as $x \to \infty$.

Therefore, taking into account the relation $U(S_n) \sim K(n)$, $n \to \infty$, and that $T(x)$ is a regularly varying function when $x \to \infty$ by using the inequality $|x + e^{-x}| \leq x^2/2$, we obtain from (2.7) that, for any sufficiently large n and for some $C_o > 0$,

$$T(U(S_n)) - T(K(n)) \leq C_o T(K(n)) g(K(n)).$$

From this inequality there follows the relation

$$T(U(S_n)) = T(K(n))(1 + o(g(K(n)))). \qquad (2.9)$$

Using similar arguments, we obtain:

$$T(n - n/\lambda_\alpha(n) + U(S_n)) = T(n)(1 + o(ng(n))), \quad n \to \infty. \qquad (2.10)$$

Since $T(K(n)) = T(n) x^{1/\alpha(n)}$, we have from (2.9) and (2.10) that

$$R_1 = \ln \frac{1 + O(ng(n))}{x^{1/\alpha(n)}(1 + O(g(K(n))))}, \quad n \to \infty.$$

Consequently, it follows from (2.6) that

$$\lim_{n \to \infty} R_1 = -\ln x. \qquad (2.11)$$

Using the monotonicity of functions $g(x)$ and $U(S)$, we have the estimate

$$R_2 \leq g^2(K(n)) C_n,$$

which shows that R_2 tends to zero as $n \to \infty$. We obtain from relations (2.4), (2,5) and (2.11) that

$$\lim_{n \to \infty} I = -\ln x. \qquad (2.12)$$

Since

$$\max_{0 \leq k \leq n} \alpha(k)(1 - S_n) \leq C_1 \alpha(n) g(K(n)),$$

for any sufficiently large n, by virtue of Lemma 1.5

$$F_n(S_n) \sim \exp\left\{ -\sum_{k=0}^{n} (1 - h_k(f_{n-k}(S_n))) \right\}.$$

Consequently, by relation (2.12) we have:
$$\lim_{n\to\infty} Me^{-g(K(n))Z(n)} = x.$$
We obtain from the last relation that $Z(n) \in B_1$, as in the proof of Theorem 1.1.

Now let $c=0$, $\nu<1$ and conditions (2.3) be fulfilled. We consider
$$I = \sum_{k=0}^{n} \alpha(k)g(n-k+U(S_n)), \quad S_n = e^{-\tau g(\psi(n))}.$$
It follows from (2.8) and (2.3) that $U(S_n) \sim \tau^{-\nu}\psi(n) = o(n)$ and, therefore, the last sum is representable in the form:
$$I = \sum_{k=0}^{n-[\varepsilon^{-1}U(S_n)]-1} \alpha(k)g(n-k+U(S_n)) +$$
$$+ \sum_{k=n-[\varepsilon^{-1}U(S_n)]}^{n} \alpha(k)g(n-k+U(S_n)) = I_1 + I_2, \quad (2.13)$$
for a sufficiently large n and any $\varepsilon>0$.

First we estimate the second term. Since $\alpha(n)$ is the regularly varying function,
$$I_2 = (1+o(1))\alpha(n) \sum_{k=0}^{[\varepsilon^{-1}U(S_n)]} g(k+U(S_n)), \quad n\to\infty.$$
Granting that $N(k+U(S_n)) = N(U(S_n)\rho(k,n))$, where $1\leq\rho(k,n)\leq 1+\varepsilon^{-1}$, when n is sufficiently large and $0\leq k\leq [\varepsilon^{-1}U(S_n)]$, using the uniform convergence theorem of slowly varying functions (Seneta (1985)) we obtain that
$$\sup_{0\leq k\leq \varepsilon^{-1}U(S_n)} |1 - \frac{N(k+U(S_n))}{N(U(S_n))}| \to 0, \quad n\to\infty.$$
Therefore,
$$I_2 = (1+o(1))\alpha(n)U(S_n)g(U(S_n)) \sum_{k=0}^{[\varepsilon^{-1}U(S_n)]} \left(1+\frac{k}{U(S_n)}\right)^{-1/\nu} \frac{1}{U(S_n)},$$
where as $n\to\infty$
$$\sum_{k=0}^{[\varepsilon^{-1}U(S_n)]} \left(1+\frac{k}{U(S_n)}\right)^{-1/\nu} (U(S_n))^{-1} \sim \int_0^{\varepsilon^{-1}} \frac{dx}{(1+x)^{1/\nu}} =$$

§3.2. Increasing Immigration

$$= \frac{\nu}{1-\nu}\left[1-\left(\frac{\varepsilon}{1+\varepsilon}\right)^{\frac{1-\nu}{\nu}}\right].$$

Then, since $g(U(S_n))=1-S_n \sim \tau g(\psi(n))$ as $n\to\infty$, granting the choice of the function $\psi(n)$, we have the equality:

$$\lim_{n\to\infty} I_2 = \frac{\nu\theta}{1-\nu}\tau^{1-\nu}\left[1-\left(\frac{\varepsilon}{1+\varepsilon}\right)^{\frac{1-\nu}{\nu}}\right]. \qquad (2.14)$$

Go over to the estimate of I_1. Let $\alpha(n)=n^\alpha l(n)$, where $\alpha\geq 0$, and $l(n)$ be a slowly varying function at infinity. If $\alpha=0$, then using the function $\lambda_\alpha(n)\to\infty$ for which

$$\sup_{n/\lambda_\alpha \leq x \leq n}\left|1-\frac{\alpha(x)}{\alpha(n)}\right| \to 0, \quad n\to\infty,$$

we shall divide I_2 into two parts:

$$I_1 = \sum_{k=0}^{[n/\lambda_\alpha(n)]} \alpha(k)g(n-k+U(S_n)) + \sum_{k=[n/\lambda_\alpha(n)]+1}^{n-[\varepsilon^{-1}U(S_n)]} \alpha(k)g(n-k+U(S_n)) =$$

$$= R_1 + R_2.$$

Since the function $g(s)$ is monotone, $R_1 \leq g\left(n - \frac{n}{\lambda_\alpha(n)}\right) \sum_{k=0}^{[n/\lambda_\alpha(n)]} \alpha(k)$. It follows from Theorem 2.1 of Seneta (1985) that

$$\sum_{k=0}^{[n/\lambda_\alpha(n)]} \alpha(k) \sim \alpha(n)n/\lambda_\alpha(n).$$

Therefore,

$$R_1 = O(g(n)\alpha(n)n/\lambda_\alpha(n)) = o(1), \quad n\to\infty.$$

By virtue of the choice of function $\lambda_\alpha(n)$, we obtain for some $0<M_0<\infty$ that

$$R_2 \leq M_0 \alpha(n) \sum_{k=[\varepsilon^{-1}U(S_n)]}^{n} g(k+U(S_n)).$$

Let $\delta>0$ be a number such that $\nu^{-1}-\delta>1$. Then, denoting $u_n=U(S_n)$, we have:

$$\sum_{k=[\varepsilon^{-1}u_n]}^{n} g(k+u_n) \leq \sup_{[\varepsilon^{-1}u_n]\leq k} \{(k+u_n)^{-\delta}N(k+u_n)\} \sum_{k=[\varepsilon^{-1}u_n]}^{\infty} (k+u_n)^{\delta-1/\nu}.$$

Using the well-known property of slowly varying functions (Seneta, 1985, p.67), we obtain:

$$\sup_{[\varepsilon^{-1}u_n] \le k} \{(k+u_n)^{-\delta} N(k+u_n)\} \sim (\varepsilon^{-1}u_n + u_n)^{-\delta} N(u_n).$$

Therefore, for any sufficiently large n and some $0 < M_1 < \infty$,

$$R_2 \le M_1 \alpha(n) U(S_n) g(U(S_n)) \left(\frac{\varepsilon}{1+\varepsilon}\right)^{\frac{1}{\nu}-1}. \tag{2.15}$$

Now let $\alpha > 0$. In this case

$$I_1 \le \sup_{0 \le k \le n} \alpha(k) \sum_{k=[\varepsilon^{-1}U(S_n)]}^{n} g(k+U(S_n)).$$

Since by virtue of property 4° (Seneta, 1985, p. 25) $\sup_{1 \le k \le n} k^\alpha l(k) \sim \alpha(n)$, we obtain the following estimate:

$$I_1 \le M_2 \alpha(n) U(S_n) g(U(S_n)) \left(\frac{\varepsilon}{1+\varepsilon}\right)^{\frac{1-\nu}{\nu}}.$$

Using that $g(U(S_n)) \sim \tau g(\psi(n))$, $U(S_n) \sim \tau^{-\nu} \psi(n)$ and $\alpha(n)g(\psi(n))\psi(n) \to \theta \in (0,\infty)$, we find that for some $0 < M_3 < \infty$

$$\limsup_{n \to \infty} I_1 \le M_3 \left(\frac{\varepsilon}{1+\varepsilon}\right)^{\frac{1-\nu}{\nu}}. \tag{2.16}$$

As $\varepsilon > 0$ is arbitrary, it follows from relations (2.13), (2.14) and (2.16) that:

$$\lim_{n \to \infty} I = \frac{\nu \theta}{1-\nu} \tau^{1-\nu}. \tag{2.17}$$

Since $g(x)$ is monotone,

$$\sum_{k=0}^{n} \beta(k) g^2(n-k+U(S_n)) \le \tau^2 g(\psi(n)) C_n.$$

This together with (2.17) and (1.9) shows that

$$\lim_{n \to \infty} \sum_{k=0}^{n} (1-h_k(f_{n-k}(S_n))) = \frac{\nu \theta}{1-\nu} \tau^{1-\nu}.$$

In order to obtain statement $Z(n) \in B_2$, it is sufficient to use Lemma 1.5, granting that $\alpha(n)(1-S_n) \to 0$ as $n \to \infty$.

Now consider the case $c \in (0, \infty)$ and $g^2(n) C_n \to 0$, $n \to \infty$. Let

§3.2. Increasing Immigration 131

$$I = \sum_{k=0}^{n} \alpha(k)g(n-k+U(S_n)), \quad S_n = e^{-\tau g(n)}.$$

It follows from (2.8) that $U(S_n) \sim \tau^{-\nu} n$, $n \to \infty$, and, since $N(x)$ is slowly varying, $N(U(S_n)) \sim N(n)$, $n \to \infty$. Consequently, by virtue of the uniform convergence theorem for slowly varying functions,

$$\sup_{0 \le k \le n} \left| 1 - \frac{N(n-k+U(S_n))}{N(n)} \right| \to 0, \quad n \to \infty. \quad (2.18)$$

Then in the following relation:

$$I = \sum_{k=0}^{[n/\lambda_N(n)]-1} \alpha(k)g(n-k+U(S_n)) + \sum_{k=[n/\lambda_N(n)]}^{n} \alpha(k)g(n-k+U(S_n)) = I_1 + I_2, \quad (2.19)$$

the first term has the estimate

$$I_1 = O(g(n)\alpha(n)n/\lambda_N(n)) = o(1). \quad (2.20)$$

Now we estimate the second term. Granting the condition $\alpha(n)g(n)n \to c$, $n \to \infty$, and using relations (1.4), (2.18), we obtain that as $n \to \infty$

$$I_2 = (1+o(1))c \sum_{k=[n/\lambda_N(n)]}^{n} k^{1/\nu - 1} (1+\pi(n,k))^{1/\nu} (n+\tau^{-\nu} n-k)^{-1/\nu},$$

where $\Pi(n,k)$ tend to zero as $n \to \infty$ uniformly with respect to $0 \le k \le n$. Hence,

$$I_2 = (1+o(1))c \sum_{k=0}^{n} \frac{(k/n)^{1/\nu - 1}}{(1+\tau^{-\nu} - k/n)^{1/\nu}} n^{-1}.$$

Since

$$\lim_{n \to \infty} \sum_{k=0}^{n} \frac{(k/n)^{1/\nu - 1}}{(1+\tau^{-\nu} - k/n)^{1/\nu}} n^{-1} = \int_{0}^{1} \frac{x^{1/\nu - 1} dx}{(1+\tau^{-\nu} - x)^{1/\nu}},$$

the last relation is equivalent to

$$\lim_{n \to \infty} I_1 = -\ln \varphi(c, \nu, \tau).$$

In order to complete the proof, it is sufficient to use relation (1.9) and Lemma 1.5. The theorem is proved.

Now we shall consider the case where $\alpha(n)g(n) \to \infty$ as $n \to \infty$. Using (2.8), it is possible to show that in this case $U(S_n)/n \to \infty$, where $S_n = \exp\{-\tau/a_2(n)\}$, $\tau > 0$, $a_2(n) = \sum_{k=0}^{n} \alpha(k)$. Then, as $n \to \infty$,

$$\sum_{k=0}^{n} \alpha(k)g(n-k+U(S_n)) \sim g(U(S_n))a_2(n) \sim \tau$$

$$\sum_{k=0}^{n} \beta(k) g^2(n-k+U(S_n)) \sim \frac{\tau^2}{a_2^2(n)} C_n.$$

Hence the following statement is valid.

Proposition 2.1. If under conditions (1.1) and (1.8) $\alpha(n)g(n) \to \infty$, $C_n = o(a_2^2(n))$, then $Z(n)/MZ(n) \to 1$ as $n \to \infty$ in probability.

The proposition shows that in this case it is possible to obtain central limit theorem type results.

§2b. The Process with Finite Variance

Up to now we have considered branching processes with immigration defined by sequences of independent and identically distributed random variables ξ_k, $k \in N_0$, (see §2.1). Now we introduce a more general scheme of immigration.

Let $\{\mu_{ij}(n), n \in N_0\}$, i, $j \in N$, be a family of independent and identically distributed Galton-Watson processes, and let $\{\eta(k,n), k=0,\ldots,n\} n \in N_0$ be a family of (generally dependent) random variables taking non-negative integer values. Assume that the variables $\eta(k,n)$ and processes $\mu_{ij}(n)$ are independent for all i, j, k, n. If we put

$$Z(0)=0, \quad Z(n)= \sum_{k=1}^{n} \sum_{j=1}^{\eta(k,n)} \mu_{jk}(n-k), \qquad (2.21)$$

then $Z(n)$ is a process with immigration defined by dependent variables.

We obtain a limit distribution for $Z(n)$, using the results of §1.3. First we consider the case $\eta(k,n) \equiv \eta(k)$. Let us denote $\mu_{11}(n) = \mu(n)$, $\alpha(n) = M\eta(n)$. It is known (see §2.1) that, if $A = M\mu(1)$ and $B = D\mu(1)$ are finite, then $a(n) = M\mu(n) = A^n$ and

$$b(n) = D\mu(n) = \begin{cases} BA^n(1-A^n)/A(1-A), & A \neq 1, \\ Bn, & A=1. \end{cases} \qquad (2.22)$$

Introduce the following notations (see §1.3):

$$A_n = \sum_{k=0}^{n} \alpha(k) A^{n-k}, \quad \Delta_n^2 = \frac{B}{(1-A)A} \sum_{k=0}^{n} \alpha(k) A^{n-k}(1-A^{n-k}),$$

$$\tilde{S}_n^{(1)} = \sum_{k=0}^{n} \eta(k) A^{n-k}, \quad \tilde{S}_n^{(2)} = \sum_{k=0}^{n} \eta(k) b(n-k), \quad \sigma_n^2 = D\tilde{S}_n^{(1)}, \quad B_n^2 = \Delta_n^2 + \sigma_n^2,$$

$$\gamma(k) = M|\eta(k) - \alpha(k)|.$$

§3.2. Increasing Immigration

Let $B(\varphi_t(y), \Pi(t,x))$ be the class of processes from Definition 0.1 and
$$\varphi_n(y)=(y-A_n)/B_n, \quad \Phi_c(x)=\Phi(x/c).$$

Corollary 2.1. Let $A<1$, $B\in(0,\infty)$, $\alpha(n)$ tends to infinity monotonically and
$$\max_{1\le k\le n} D\eta(k) = o(B_n^4), \quad n\to\infty.$$

$1°$ If $B_n^2 \sim \Delta_n^2$, then $Z(n)\in B(\varphi_n(y), \Phi(x))$;

$2°$ if $B_n^2 \sim \Delta_n^2$, $\widetilde{S}_n^{(1)} \in B(\varphi_n(y), V(x))$, then $Z(n)\in B(\varphi_n(y), V(x))$;

$3°$ if $B_n^2 \sim \theta^{-2}\sigma_n^2$, $0<\theta<1$, $\widetilde{S}_n^{(1)}\in B((y-A_n)/\sigma_n, V(x))$, then
$$Z(n)\in B(\varphi_n(y), \Phi_{1-\theta}*V_\theta(x)), \quad V_\theta(x)=V(x/\theta).$$

Proof. We shall prove that the conditions of Theorem 1.3.1 are fulfilled. Let us denote $\hat{\Phi}_{kn}(t)=M\exp\{it(\mu_{jk}(n-k)-A^{n-k})\}$. Using Chebyshev's inequality and the estimate
$$|1-\hat{\Phi}_{kn}(t)| \le t^2 b(n-k)/B_n^2,$$
we obtain that for any $\varepsilon>0$
$$P\left\{\sum_{k=1}^n \eta(k) \left|1-\hat{\Phi}_{kn}\left(\frac{t}{B_n}\right)\right|^2 > \varepsilon\right\} \le \frac{t^4}{B_n^4 \varepsilon} \sum_{k=1}^n \alpha(k) b^2(n-k).$$

Since $B_n\to\infty$, $n\to\infty$, we have from this:
$$\sum_{k=1}^n \eta(k) \left|1-\hat{\Phi}_{kn}\left(\frac{t}{B_n}\right)\right|^2 \xrightarrow{P} 0. \qquad (2.23)$$

If we use Chebyshev's inequality again, we obtain for any $\varepsilon>0$ that
$$P\left\{\frac{1}{B_n^2}|\widetilde{S}_n^{(2)}-\Delta_n^2|>\varepsilon\right\} \le \frac{C}{B_n^2} \max_{1\le k\le n}\sqrt{D\eta(k)}.$$

Hence, under our conditions,
$$\left(\widetilde{S}_n^{(2)}-\Delta_n^2\right)/B_n^2 \xrightarrow{P} 0, \quad n\to\infty. \qquad (2.24)$$

Now we prove that as $n\to\infty$
$$L_n(\varepsilon)= \sum_{k=1}^n \eta(k) M\left[\left(\frac{\mu(n-k)-A^{n-k}}{B_n}\right)^2; |\mu(n-k)-A^{n-k}|>\varepsilon B_n\right] \xrightarrow{P} 0.$$

It is easy to show that $L_n(\varepsilon)-ML_n(\varepsilon)\xrightarrow{P} 0$, $n\to\infty$. Therefore we must consider $ML_n(\varepsilon)$. We have for any integer $N>0$

$$ML_n(\varepsilon) \le \frac{1}{\Delta_n^2} \sum_{k=0}^{N} \alpha(n-k) M[(\mu(k)-A^k)^2; |\mu(k)-A^k|>\varepsilon\Delta_n]$$
$$+ \frac{1}{\Delta_n^2} \sum_{k=N}^{n} \alpha(n-k) M[(\mu(k)-A^k)^2; |\mu(k)-A^k|>\varepsilon\Delta_n]. \quad (2.25)$$

Since $\alpha(n)$ is monotone, the first term in (2.25) has the estimate

$$\text{const} \cdot \sum_{k=0}^{N} M[(\mu(k)-A^k)^2; |\mu(k)-A^k|>\varepsilon\Delta_n],$$

that tends to zero as $n \to \infty$ for any fixed N. Using the monotonicity of $\alpha(n)$ again, we obtain that the second term is not greater than $A^{N-1}(1-A)^{-2}$. This estimation shows that the second term is small for a sufficiently large N. Hence $L_n(\varepsilon) \xrightarrow{P} 0$, $n \to \infty$.

In order to obtain the statement of the corollary, we shall use Theorem 1.3.1. Let N_n be a sequence of integers such that

$$P\{\max_{1 \le k \le n} \eta(k) \ge N_n\} \to 0, \quad n \to \infty.$$

Putting

$$\nu((k-1)N_n+j,n) = \chi(j \le \eta(k)), \quad X((k-1)N_n+j,n) = \mu_{jk}(n-k),$$

where $j=1,\ldots,N_n$, $k=1,\ldots,n$, we can see that the conditions of Theorem 1.3.1 are fulfilled. Therefore Corollary 1.3.2 is applicable to (2.21). The corollary is proved.

If $D = M\mu^3(1) < \infty$, then it is easy to calculate that

$$C_3(k) = M|\mu(k)-A^k|^3 = \frac{A^{3k}C_1 - A^{2k}C_2 + A^k C_3}{(A^3-A)(A^2-A)}, \quad A \ne 1,$$

where C_i are some positive constants depending on A, B and D.

Now we present a corollary of Theorem 1.3.2. Put

$$V_n^*(x) = P\left\{\frac{\tilde{S}_n^{(1)} - A_n}{B_n} < x\right\}.$$

Corollary 2.2. If $A \ne 1$, $D < \infty$, then there exists $C > 0$ such that for $n \ge 1$

$$\sup_x \left| P\left\{\frac{Z(n)-A_n}{B_n} < x\right\} - \Phi_{\Delta_n/B_n} * V_n^*(x) \right| \le$$

$$\le C \left[\frac{1}{\Delta_n^3} \sum_{k=0}^{n} \alpha(k) C_3(n-k) + \hat{\gamma}(n) \right],$$

where

§3.2. Increasing Immigration

$$\hat{\gamma}(n) = \left[\sum_{k=0}^{n} \alpha(k) A^{n-k}(1-A^{n-k})\right]^{-1} \sum_{k=0}^{n} \gamma(k) A^{n-k}(1-A^{n-k}).$$

We can obtain the rate of convergence in the statements of Corollary 2.1 with the help of Corollary 2.2. In addition it is possible to deduce limit theorems for supercritical and critical processes from Theorem 1.3.1.

The results of §1.3 can be used for the investigation of the total progeny of the branching process. In fact, it is easy to see that the variable $Y(n) = Z(0) + Z(1) + \ldots + Z(n)$ is representable in the form

$$Y(n) = \sum_{k=0}^{n} \sum_{i=1}^{\eta(k)} X_{ki}(n-k) = \sum_{k=0}^{n} \sum_{i=1}^{\eta(k)} \sum_{j=0}^{n-k} \mu_{ki}(j),$$

where $X_{ki}(n-k)-1$ is the total number of particles generated by the ith particle immigrating at time k. Since the particles evolve independently of each other, variables $X_{ki}(n)$ are independent and have the same distribution for different k and i. Thus, we can obtain limit theorems for $Y(n)$ under different conditions for immigration, using Theorem 1.3.1.

Let $Z(n)$ be a process with immigration defined by the family of random variables $\{\nu(k,n), k=1,\ldots,n\}$, $n \in N_o$. If we suppose $\nu(n,0) = \alpha(n)$, $\nu(n,k) = 0$, $k \geq 1$, where $\alpha(n)$, $n \geq 0$, is an integer-valued function, then $Z(n)$ is a branching process starting from increasing number of ancestors considered, for example, by Lamperti (1967a) and by R. Ibragimov (1972).

It is obvious that in this case

$$A_n = \alpha(n) A^n, \quad \sigma_n^2 = 0, \quad B_n^2 = \alpha(n) D\mu(n).$$

Using the fact that $\mu(n)/\sqrt{D\mu(n)} \to \mu$ with probability 1 when $A > 1$, $B < \infty$, where $M\mu = 1$, $D\mu < \infty$ (see Rahimov and Sirazhitdinov, 1984), one can show that

$$L_n(\varepsilon) = \frac{1}{D\mu(n)} M[(\mu(n) - A^n)^2; \ |\mu(n) - A^n| > \varepsilon B_n] \to 0$$

as $n \to \infty$. Consequently, we obtain from Theorem 1.3.1 the following result of Lamperti (1967a).

Corollary 2.3. If $A > 1$, $B < \infty$ and $\alpha(n) \to \infty$ as $n \to \infty$, then

$$\lim_{n \to \infty} P\{(\mu(n) - A_n)/B_n < x \mid \mu(0) = \alpha(n)\} = \Phi(x).$$

In addition we can find from Corollary 1.3.2 the following result of R. Ibragimov (1972).

Corollary 2.4. Let $A > 1$, $D < \infty$, and let δ_n be the uniform error of limiting distribution from $\Phi(x)$ in Corollary 2.3. Then, for some positive constant C and $n \geq 1$

$$\delta_n \leq CM|\mu(n)-A^n|^3 / \sqrt{\alpha(n)} \, (D\mu(n))^{3/2}.$$

It is shown by R. Ibragimov (1972) that

$$\frac{M|\mu(n)-A^n|^3}{(D\mu(n))^{3/2}} \leq \frac{A^2(3AB^2+10A^3B+2A^2D+4A^5)}{(A^3-A)(A-1)(D\mu(1))^{3/2}}, \quad n \geq 1.$$

§3.3. LOCAL LIMIT THEOREMS

§3a. Occupation of an Increasing State

The local probabilities of branching processes are less investigated than integral properties of the number of particles. At present, local limit theorems have actually been proved only for Markov branching processes without immigration (Chistyakov (1957), Kesten, Ney and Spitzer (1966), Topchii (1982)) or for processes with homogeneous immigration (Mellein (1982)). These results are the local analogy of classical limit theorems of the theory of branching processes. On the asymptotic properties of an invariant measure, the Green and harmonic functions of Galton-Watson processes, the reader can refer to Chapter 2 of the book of Athreya and Ney (1972) (see also Vatutin and Zubkov (1985) for details).

In the present section, using the methods of §3.1, we shall prove local limit theorems in the "non-classical" situation. More exactly speaking, the local limit theorems for the process $Z(n)$ considered in §3.1 will be proved in the case where $M\xi_n \to 0$ as $n \to \infty$, and is a regularly varying function. We will find that, under some conditions on parameters, probabilities

$$P_{ij}(n) = P\{Z(n)=j \mid Z(0)=i\}$$

(i is fixed) can have two different asymptotes, depending on the relation of the rate of convergence of j and n to infinity, as in the integral limit theorems (see Corollary 3.1).

Let $\alpha(n)$, $\beta(n)$, $\theta_1(n)$, $\mu(n)$, $f_n(s)$, $f(s)=f_1(s)$ be the as in §3.1, and let

$$A=f'(1), \; B=f''(1), \; \gamma_n=2\alpha(n)/B, \; Q_{ij}(n)=P\{\mu(n)=j|\mu(0)=i\}.$$

In this section we shall assume that the process $\mu(n)$ is critical and

§3.3. Local Limit Theorems

non-periodical, that is, A=1 and the greatest common divisor of $\{k: Q_k(1)>0\}=1$, $Q_1(1)\neq 1$, and condition (3.1.8) is fulfilled. In addition assume that

$$\sum_k k^2 Q_k(1) \ln k < \infty, \qquad (3.1)$$

that $\alpha(n)$ is a slowly varying function as $n\to\infty$ and that

$$\lim_{n\to\infty} |\alpha(n\varepsilon)-\alpha(n)| \ln n = 0 \qquad (3.2)$$

for any $\varepsilon\in(0,1)$.

Definition 3.1. We shall say that the function $\varphi(x,y)$ belongs to the class $\mathcal{K}(a,b)$, if $a<\varphi(x,y)<b$ and

$$a < \liminf_{\substack{x\to\infty \\ y\to\infty}} \varphi(x,y) \leq \limsup_{\substack{x\to\infty \\ y\to\infty}} \varphi(x,y) < b.$$

We will use the notations

$$K_1=\{\theta_1(n): \liminf_{n\to\infty} \theta_1(n)>0\}, \quad K_2=\{\theta_1(n): \limsup_{n\to\infty} \theta_1(n)<\infty\}.$$

Theorem 3.1. Let conditions (3.1), (3.2) be satisfied and $\liminf\limits_{n\to\infty} n^{\alpha(n)}>1$. If $\alpha(n)\to 0$, $\beta(n)\to 0$, $j\to\infty$ and $n\to\infty$ such that $(j/n)^{\gamma_n}\in\mathcal{K}(0,1)$, then for any fixed $i\in N_0$

$$\frac{Bj}{2\alpha(n)}(j/n)^{-\gamma_n} P_{ij}(n) \to 1.$$

Now we mention theorems for the case where $\alpha(n)$ is a regularly varying function and

$$\lim_{n\to\infty} n^{\alpha(n)} = 1. \qquad (3.3)$$

Theorem 3.2. Let conditions (3.1) and (3.2) be satisfied. If $\theta_1(n)\in K_1$, $\beta(n)=o(\alpha(n))$, n and $j\to\infty$ such that $\frac{\ln j}{\ln n} \in \mathcal{K}(0,1)$, then $\frac{j}{\gamma_n} P_{ij}(n) \to 0$, $i\in N_0$.

Theorem 3.3. Let conditions (3.1) and (3.3) be satisfied. If $\theta_1(n)\in K_2$, $\beta(n)=o(\alpha(n))$, n, $j\to\infty$ such that $j/n\in\mathcal{K}(0,C)$, where $C\in(0,\infty)$, then

$$\left(\sum_{k=0}^n \alpha(k)+i\right)^{-1} \left(\frac{Bn}{2}\right)^2 e^{\frac{2j}{Bn}} P_{ij}(n) \to 1, \quad i\in N_0.$$

Corollary 3.1. Let conditions (3.1), (3.3) be satisfied and $\theta_1(n)\in K_1\cap K_2$.

1° If n, $j\to\infty$, $\ln j/\ln n\in\mathcal{K}(0,1)$, then $P_{ij}(n)\sim\frac{2\alpha(n)}{Bj}$;

2° if n, $j\to\infty$, $j/n\in\mathcal{K}(0,C)$, $C\in(0,\infty)$, then

$$P_{ij}(n) \sim \left(\sum_{k=0}^{n} \alpha(k)+i \right) \left(\frac{2}{Bn} \right)^2 e^{-\frac{2j}{Bn}}.$$

The proofs of theorems 3.2 and 3.3 are different from the proof of Theorem 3.1. They use the same scheme as the proof of Theorem 1.1. Namely, first we prove the local theorems for some partial processes and then the theorems for $Z(n)$ will be obtained from a representation of its local probabilities by corresponding probabilities of the partial processes.

It is known (Kesten et. al.(1966)) that under the condition (3.1) the relation

$$\lim_{j \to \infty} G(i,j) j = 2i/B, \quad i \in N_o, \tag{3.4}$$

is valid for Green's function $G(i,j)$ defined in §3.1.

We put $D=[s: |s| \leq 1]$. Define an L_1 norm of power series of the form $a(s) = \Sigma_k a_k s^k$ that converge absolutely on the domain D as:

$$\|a(s)\|_{N_1}^{N_2} = \sum_{k=N_1}^{N_2} |a_k|, \quad \|a(s)\| = \|a(s)\|_0^\infty,$$

$N_1, N_2, \ldots$ here and later on denote different positive integers. We put also $[a(s)]_j = |a_j|$. Then we can see that

$$\sup_{s \in D} |a(s)| \leq \|a(s)\|, \quad \|a(s)b(s)\| \leq \|a(s)\| \cdot \|b(s)\|.$$

If $a_i^{(k)}(s) = \sum_{j=0}^{\infty} a_{ij}^{(k)} s^j$, $k=1,2$; $i=1,\ldots,m$, $m \in N$, and the relation

$$a(s) = \sum_{i=1}^{m} a_i^{(1)}(s) a_i^{(2)}(s)$$

is true, then

$$\sup_{N_1 \leq j \leq N_2} [a(s)]_j \leq \sum_{i=1}^{m} \|a_i^{(1)}(s)\|_{N_1}^{N_2} \sup_{N_1 \leq j \leq N_2} [a_i^{(2)}(s)]_j. \tag{3.5}$$

Note that a particular form $(N_1=0, N_2=\infty)$ of the norm $\|\cdot\|_{N_1}^{N_2}$ was used by Kesten et. al. (1966).

Lemma 3.1. If conditions (3.1), (3.2) are satisfied and $\alpha(n) \to 0$, $\beta(n) \to 0$, then there exists a positive integer N, such that

$$F_o(0,n,s) = \hat{F}_N(n,s) \left[\frac{1+(N-1)2^{-1}B(1-s)}{1+2^{-1}Bn(1-s)} \right]^{\gamma_n} \prod_{k=N}^{n} (1+g_n(k,s)), \tag{3.6}$$

§3.3. Local Limit Theorems

where

$$\limsup_{n\to\infty, s\in D} |1-\hat{F}_N(n,s)|=0, \quad \lim_{n\to\infty} \sum_{k=N}^{n} \|g_n(k,s)\|=0.$$

For the proof of this lemma readers can see Rahimov's paper (1988a).

Lemma 3.2. Let $m\to\infty$ and $i\to\infty$ such that i/m is bounded and condition (3.1) is satisfied. Then, for any N_0,

$$\sum_{k=N_0}^{n} P\{\mu(k)=i \mid \mu(0)=1\} \sim \frac{2}{Bi}\left(1-\frac{2}{Bm}\right)^i.$$

Proof. Let $\limsup m/i<\infty$. It is known (Kesten et.al., 1966) that under the condition (3.1) it is possible to choose an integer N that for $S\in D$ and $k\geq N$

$$f_k(s)-f_k(0)=(1-f_k(0))\frac{1+(N-1)\alpha-(N-1)\alpha s}{1+(k-1)\alpha-(k-1)\alpha s} A(k,N,s), \quad (3.7)$$

where $\alpha=B/2$, and $A(k,N,s)=\sum_{j=1}^{\infty} r_j(k,N)s^j$ converges uniformly for $s\in D$ and for some $M<\infty$

$$\|A(k,N,s)\|\leq M, \quad k\geq N. \quad (3.8)$$

We consider

$$\sum_{i=1}^{\infty}\sum_{k=N}^{m} Q_i(k)s^i = \sum_{k=N}^{m} [f_k(s)-f_k(0)]. \quad (3.9)$$

Using (3.7), we obtain from (3.8) that for each $i\in N$

$$\sum_{k=N}^{m} Q_i(k)=(1+(N-1)\alpha)\sum_{k=N}^{m} \frac{1-f_k(0)}{1+(k-1)\alpha} r_i(k,N) +$$

$$+ \sum_{k=N}^{m} \frac{(1-f_k(0))(k-N)}{(1+(k-1)\alpha)(k-1)} \sum_{j=1}^{i-1} r_j(k,N)\left[\frac{(k-1)\alpha}{1+(k-1)\alpha}\right]^{i-j} =I_1+I_2. \quad (3.10)$$

Since

$$sA'(k,N,s)(1-f_k(0)) = \frac{sf'_k(s)(1+(k-1)\alpha-(k-1)\alpha s)}{1+(N-1)\alpha-(N-1)\alpha s} -$$

$$- \frac{(f_k(s)-f_k(0))(k-N)\alpha s}{(1+(N-1)\alpha-(N-1)\alpha s)^2},$$

using (3.5), we have for any $2\leq N_1<N_2<\infty$:

$$\sup_{N_1 \le j \le N_2} |jr_j(k,N)| \le \frac{k-N}{(1-f_k(0))(N-1)} \left[\frac{(N-1)\alpha}{1+(N-1)\alpha}\right]^{N_1} \sup_{j \ge 1} jQ_j(k). \qquad (3.11)$$

Consider the first term in (3.10). Taking into account estimation (3.11), we have for any $i \ge N_1 \ge 2$ the inequality

$$|iI_1| \le \frac{1+\alpha}{\alpha} \left[\frac{(N-1)\alpha}{1+(N-1)\alpha}\right]^{N_1} \sup_{j \ge 1} jG(1,j).$$

The right-hand side of this inequality will be small for any sufficiently large N_1. Therefore, when $i \to \infty$,

$$I_1 = o(i^{-1}). \qquad (3.12)$$

Let $L(m)$ be an integer-valued function such that $L(m) \to \infty$, $L(m)/m \to 0$ as $m \to \infty$, but $i/L(m) \to \infty$, $i/L(m)\ln i \to 0$. Consider the relation

$$I_2 = \sum_{k=N}^{L(m)} \frac{(1-f_k(0))(k-N)}{(1+(k-1)\alpha)(k-1)} \sum_{j=1}^{[i/\ln i]} r_j(k,n) \left[\frac{(k-1)\alpha}{1+(k-1)\alpha}\right]^{i-j} +$$

$$+ \sum_{k=N}^{m} \frac{(1-f_k(0))(k-N)}{(1+(k-1)\alpha)(k-1)} \sum_{j=[i/\ln i]+1}^{i-1} r_j(k,N) \left[\frac{(k-1)\alpha}{1+(k-1)\alpha}\right]^{i-j} +$$

$$+ \sum_{k=L(m)+1}^{m} \frac{(1-f_k(0))(k-N)}{(1+(k-1)\alpha)(k-1)} \sum_{j=1}^{[i/\ln i]} r_j(k,N) \left[\frac{(k-1)\alpha}{1+(k-1)\alpha}\right]^{i-j} +$$

$$+ R_1 + R_2 + R_3. \qquad (3.13)$$

If we use (3.8), then we have the following estimate for the first term

$$|R_1| \le c_4 \sum_{k=N}^{L(m)} (1+(k-1)\alpha)^{-2} \left[\frac{(k-1)\alpha}{1+(k-1)\alpha}\right]^{i-[i/\ln i]} \le$$

$$\le c_5 i^{-1} \left(1 - \frac{1}{\alpha L(m)}\right)^i = o(i^{-1}), \quad i,m \to \infty. \qquad (3.14)$$

Using (3.4) and (3.11) with $N_1 = [i/\ln i]$, $N_2 = i-1$, we obtain:

$$R_2 = o(i^{-1}), \quad i,m \to \infty. \qquad (3.15)$$

Finally we consider R_3. It follows from (3.8) that

§3.3. Local Limit Theorems

$$\lim_{i \to \infty} \sum_{j=[i/\ln i]}^{\infty} |r_j(k,N)| = 0.$$

Then, when $i \to \infty$

$$\hat{R}_3 = \sum_{k=L(m)+1}^{m} \frac{1-f_k(0)}{1+(k-1)\alpha} \left(1 - \frac{1}{1+(k-1)\alpha}\right)^i \sum_{j=1}^{[i/\ln i]} r_j(k,N) =$$

$$= (1+o(1)) \sum_{k=L(m)+1}^{m} \frac{(1-f_k(0))}{(1+(k-1)\alpha)} \left(1 - \frac{1}{1+(k-1)\alpha}\right)^i. \qquad (3.16)$$

Since the fraction in the brackets in the relation $R_3 = \hat{R}_3(1+(R_3-\hat{R}_3)/\hat{R}_3)$ tends to zero when i and $m \to \infty$ by virtue of the property of $L(t)$ and (3.16), we obtain the equality

$$R_3 = (1+o(1)) \sum_{k=L(m)+1}^{m} (1+(k-1)\alpha)^{-2}(1-1/(1+(k-1)\alpha))^i =$$

$$= \frac{2}{Bi}(1 - \frac{2}{Bm})^i (1+o(1)). \qquad (3.17)$$

It follows from (3.10), (3.12)-(3.15) and (3.17) that when m, $i \to \infty$, $\limsup(m/i) < \infty$

$$\sum_{k=N}^{m} Q_i(k) \sim \frac{2}{Bi}(1 - \frac{2}{Bm})^i. \qquad (3.18)$$

Let now $i/m \to 0$. Then in the equality

$$\sum_{k=N}^{m} Q_i(k) = G(1,i)[1 - \sum_{k=m+1}^{\infty} Q_i(k)/G(1,i) - \sum_{k=0}^{N-1} Q_i(k)/G(1,i)]$$

the first fraction tends to zero, in view of the fact that

$$\sum_{k=m}^{\infty} Q_i(k)/G(1,i) \le c_6 \frac{i}{m}$$

for some c_6 and sufficiently large i,m. It is known (Kesten et. al., 1966, p.609) that

$$\sum_{k=0}^{K} Q_i(k) \le \sum_{k=0}^{[\varepsilon_i]} Q_i(k) \le c_7 \left[\frac{\ln i}{i^{3/2}} + \frac{\sqrt{\varepsilon}}{i}\right]$$

for any K, $\varepsilon > 0$ and sufficiently large i. Therefore

$$\sum_{k=0}^{K} Q_i(k) = o(i^{-1}). \qquad (3.19)$$

Thus we again have the relation (3.18). Using (3.19) once again, we obtain from (3.18) that

$$\sum_{k=N_o}^{m} Q_i(k) \sim \frac{2}{Bi}\left(1-\frac{2}{Bm}\right)^i.$$

The lemma is proved.

Lemma 3.3. If $\gamma \to 0$, $k \to \infty$, then

$1^\circ.$ $\binom{\gamma+k-1}{k} = \gamma k^{\gamma-1}(1+o(k^{-1}));$

$2^\circ.$ $\sum_{i=0}^{n} \binom{\gamma+i-1}{i} = n^\gamma(1+o(1)).$

We obtain the proof of this lemma from the following relations:

$$\binom{\gamma+k-1}{k} = \gamma \Gamma(\gamma+k)/\ \Gamma(\gamma+1)\Gamma(k-1)k(k-1), \qquad (3.20)$$

$$\sum_{i=0}^{\infty} \binom{\gamma+i-1}{i} s^i = (1-s)^{-\gamma}. \qquad (3.21)$$

We put

$$\hat{T}_n(N,s) = \hat{F}_N(n,s)(1+(N-1)2^{-1}B(1-s))^{\gamma_n} \prod_{k=N}^{n} (1+g_n(k,s)) = \sum_{j=o}^{\infty} R_j(n,N)s^j.$$

Lemma 3.4. If the conditions of Lemma 3.1 are satisfied, then there exists $M_1 < \infty$ such that for any $\varepsilon > 0$ and sufficiently large n

$$\sup_{N_1 \leq j \leq N_2} |jR_j(n,N)| \leq \varepsilon\alpha(n) + M_1 \left[\frac{N_2}{n} + \frac{1}{N_1} \sup_{N \leq k \leq n} \beta(k) + \right.$$

$$\left. \alpha(n) \sup_{N_1 \leq j \leq N_2} |1-2(Bj \sum_{k=N}^{n} Q_j(k))^{-1}(1-\frac{2}{Bn})^j| \right].$$

Proof. It follows from Lemma 3.1 that

$$T_n(N,s) = (1+\frac{Bn}{2}(1-s))^{\gamma_n} F_o(0,n,s). \qquad (3.22)$$

Then

§3.3. Local Limit Theorems

$$ST'_n(N,s) = T_n(N,s) \sum_{k=N}^{n} \frac{\partial h_{n-k}(f_k(s))}{\partial f_k(s)} s \frac{\partial f_k(s)}{\partial s} \left(\frac{1}{h_{n-k}(f_k(s))} - 1 \right) +$$

$$+ T_n(N,s) \sum_{k=N}^{n} \left(\frac{\partial h_{n-k}(f_k(s))}{\partial f_k(s)} - \alpha(n-k) \right) \frac{\partial f_k(s)}{\partial s} s +$$

$$+ T_n(N,s) \sum_{k=N}^{n} (\alpha(n-k) - \alpha(n)) \frac{\partial f_k(s)}{\partial s} s +$$

$$+ T_n(N,s) \left[\sum_{k=n}^{n} s \frac{\partial f_k(s)}{\partial s} - \frac{2^{-1} Bns}{1 + \frac{Bn}{2}(1-s)} \right] \alpha(n) = I_1 + I_2 + I_3 + I_4. \qquad (3.23)$$

We obtain from the definition of $T_n(N,s)$ that there exists M_0 such that

$$\|T_n(N,s)\| \leq M_0 < \infty. \qquad (3.24)$$

Accounting for (3.24) and using inequality (3.5), we obtain:

$$\sup_{N_1 \leq j \leq N_2} [I_1]_j \leq M_0 \sum_{k=N}^{n} \alpha(n-k) \left\| \frac{1}{h_{n-k}(f_k(s))} - 1 \right\| \sup_{N_1 \leq j \leq N_2} jQ_j(k) \leq$$

$$\leq c_8 \sum_{k=N}^{[n-n/\lambda_\alpha]} \alpha^2(n-k) \sup_{1 \leq j < \infty} jQ_j(k) + c_8 N_2 \sup_{j,k} k^2 Q_j(k) \sum_{k=[n-n/\lambda_\alpha]}^{n} \alpha^2(n-k) k^{-2},$$

where (see [45]) $\sup_{j,k} k^2 Q_j(k) < \infty$ and $\lambda_\alpha = \lambda_\alpha(n)$ is the function defined in §3.1 (see (1.38)). We obtain from this fact that for a sufficiently large n and for any $\varepsilon > 0$

$$\sup_{N_1 \leq j \leq N_2} [I_1]_j \leq \varepsilon \alpha(n) + c_9 N_2 n^{-1}. \qquad (3.25)$$

Since

$$\left\| \frac{\partial h_{n-k}(f_k(s))}{\partial f_k(s)} - \alpha(n-k) \right\|_{N_1}^{N_2} \leq \beta(n-k) \|1 - f_k(s)\|_{N_1}^{N_2} \leq \beta(n-k) N_1^{-1},$$

we have the following estimate for some c_{10}:

$$\sup_{N_1 \leq j \leq N_2} [I_2]_j \leq \frac{c_{10}}{N_1} \left[\sup_{n/\lambda_\alpha \leq k \leq n} \beta(k) + \frac{N_2}{n} \right]. \qquad (3.26)$$

We obtain, as in the estimation of $[I_1]_j$, that for a sufficiently large n and for any $\varepsilon>0$

$$\sup_{N_1 \le j \le N_2} [I_3]_j \le \varepsilon\alpha(n) + c_{11}N_2 n^{-1}. \qquad (3.27)$$

Now let us consider I_4. It is easy to see that

$$\sup_{N_1 \le j \le N_2} [I_4]_j \le c_{12}\alpha(n) \sup_{N_1 \le j \le N_2} |1 - \frac{2}{B}(j \sum_{k=n}^{n} Q_j(k))^{-1}(1-2/Bn)^j|. \qquad (3.28)$$

The statement of Lemma 3.4 follows from relations (3.23) and (3.25)-(3.28).

Proof of Theorem 3.1. Using equality (3.6), we have:

$$P_{oj}(n) = \sum_{i=0}^{j} R_i(n,N) \binom{\gamma_n+j-i-1}{j-1} \left(\frac{Bn/2}{1+Bn/2}\right)^{j-i} \left(1+Bn/2\right)^{-\gamma_n}. \qquad (3.29)$$

It follows from Theorem 1 of Rahimov's paper (1986a) that

$$\lim_{n \to \infty} T_n(N,s) = 1 \qquad (3.30)$$

uniformly on compact subsets of the domain D. Using part 1° of Lemma 3.3 and granting (3.30), we can see that the first term in (3.29) multiplied by $j\gamma_n^{-1}(j/n)^{\gamma_n}$ tends to 1 when $j, n \to \infty$ and $(j/n)^{\gamma_n} \in K(0,1)$.

Now we consider the sum

$$I = \sum_{i=1}^{j} R_i(n,N) b_{ij}^{(n)}, \quad b_{ij}^{(n)} = \binom{\gamma_n+j-i-1}{j-1} \gamma_n^{-1} j^{1-\gamma_n}.$$

Since $b_{ij}^{(n)}$ is increases with respect to i when $\gamma_n < 1$ and $b_{j/2,j}^{(n)} \sim 2$ as $j, n \to \infty$, there exists M_2 such that $|b_{ij}^{(n)}| \le M_2$, $0 \le i \le j/2$. Therefore, using (3.24) and (3.30), we have:

$$\lim_{j,n \to \infty} \sum_{i=1}^{[j/2]} R_i(n,N) b_{ij}^{(n)} = 0. \qquad (3.31)$$

On the other hand,

$$|\sum_{i=[j/2]}^{j} R_i(n,N) b_{ij}^{(n)}| \le 2\gamma_n^{-1} j^{-\gamma_n} \sup_{j/2 \le i \le j} |iR_i(n,N)| \sum_{i=0}^{j} \binom{\gamma_n+i-1}{i}.$$

By Lemma 3.4 and part 2° of Lemma 3.3, there exists M_3 such that for any $\varepsilon>0$ and for any sufficiently large n

§3.3. Local Limit Theorems

$$\left| \sum_{i=[j/2]}^{j} R_i(n,N) b_{ij}^{(n)} \right| \leq M_3 \frac{1}{\alpha(n)} \left[\varepsilon\alpha(n) + \frac{j}{n} + \frac{1}{j} \sup_{n/\lambda_\alpha \leq k \leq n} \beta(k) + \right.$$

$$\left. + \alpha(n) \sup_{j/2 \leq i \leq j} \left| 1 - \frac{2}{B} \left(i \sum_{k=N}^{n} Q_i(k) \right)^{-1} (1-2/Bn)^i \right| \right].$$

Let

$$0 < x_1 = \liminf (j/n)^{\gamma_n} \leq \limsup (j/n)^{\gamma_n} = x_2 < 1.$$

Then as $j, n \to \infty$ and $(j/n)^{\gamma_n} \in \mathcal{K}(0,1)$ we have:

$$\limsup (j/n\alpha(n)) \leq \limsup \left(x_2^{1/\gamma_n} \frac{1}{\alpha(n)} \right) = 0,$$

$$\limsup (j/j\alpha(n)) \leq \limsup \left(x_1^{1/\gamma_n} \frac{1}{n\alpha(n)} \right) = 0.$$

On the other hand, by virtue of Lemma 3.2,

$$\lim_{j,n\to\infty} \sup_{j/2 \leq i \leq j} \left| 1 - 2/B \left(i \sum_{k=N}^{n} Q_i(k) \right)^{-1} (1-2/Bn)^i \right| = 0.$$

Therefore the inequality

$$\sum_{i=[j/2]}^{j} |R_i(n,N) b_{ij}^{(n)}| < \varepsilon$$

is true for any $\varepsilon > 0$ and for any sufficiently large j and n. Consequently, it follows from this fact and from (3.31) that

$$\lim_{\substack{j,n\to\infty \\ (j/n)^{\gamma_n} \in \mathcal{K}(0,1)}} \sum_{i=1}^{j} R_i(n,N) b_{ij}^{(n)} = 0.$$

Since when $(j/n)^{\gamma_n} \in \mathcal{K}(0,1)$

$$\lim_{n\to\infty} \left(\frac{n}{1+Bn/2} \right)^{\gamma_n} = 1, \quad \lim_{j,n\to\infty} \sup_{j/2 \leq i \leq j} \left| 1 - \left(\frac{2^{-1} Bn}{1+Bn/2} \right)^{j-1} \right| = 0,$$

we obtain:

$$i/\gamma_n (j/n)^{-\gamma_n} P_{oj}(n) \to 1 \qquad (3.32)$$

as $j, n \to \infty$ such that $(j/n)^{\gamma_n} \in \mathcal{K}(0,1)$.

In the case $Z(0) = i$, with fixed i, using equality

III. Time-Dependent Immigration

$$P_{ij}(n) = \sum_{k=0}^{j} P_{ok}(n) Q_{ij-k}(n)$$

and the fact that (Kesten et. al., 1966, p. 604)

$$\sup_{r\geq 1} Q_{ir}(n) \leq iM_4/n^2,$$

we can see that (3.32) is true with $P_{ij}(n)$ instead of $P_{oj}(n)$. The theorem is thus proved.

Let us set

$$Z(N_1,N_2,n) = \sum_{k=N_1}^{N_2} \sum_{i=1}^{\xi_k} \mu_{ik}(n-k)$$

for any $0 \leq N_1 \leq N_2 \leq n-1$, where $\{\mu_{ik}(n)\}$, $i, k \in N$, are independent branching processes with the same generating function $f(s)$,

$$H_n(N_1,N_2,s) = Ms^{Z(N_1,N_2,n)}, \quad \Delta_n(N_1,N_2,s) = H_n(N_1,N_2,s) - H_n(N_1,N_2,0),$$

$$\delta_n(s) = f_n(s) - f_n(0), \quad d=[0,1].$$

Let

$$A_1(k,n,s) = \prod_{j=N_1}^{k-1} h_j(f_{n-j}(s)) \prod_{j=k+1}^{N_2} h_j(f_{n-j}(0)),$$

$$A_2(k,n,s) = \sum_{m=1}^{\infty} \sum_{i=1}^{\infty} \sum_{l=1}^{i} f_{n-k}^{i-1}(0) P\left\{\sum_{j=1}^{i-1} \mu_{jk}(n-k)=m\right\} q_i(k) s^m,$$

$$B(k,n,s) = A_1(k,n,s)[A_2(k,n,s) + \alpha(k) - \sum_{i=1}^{\infty}(1-f_{n-k}^{i-1}(0)) i q_i(k))].$$

Lemma 3.5. If conditions (3.1) and (3.3) are satisfied and $N_1 = N_1(n)$, $N_2 = N_2(n)$, then

$$\Delta_n(N_1,N_2,s) = \sum_{k=N_1}^{N_2} \delta_{n-k}(s) B(k,n,s), \qquad (3.33)$$

where, if

$$\lim_{n\to\infty} \sum_{k=N_1}^{N_2} \alpha(k)(n-k)^{-1} = 0, \text{ then } \lim_{n\to\infty} \sup_{s\in D} \sup_{N_1 \leq k \leq N_2} |1-A_1(k,n,s)| = 0,$$

$$a_o^{(2)}(k,n) = 0, \quad a_i^{(2)}(k,n) \leq \sum_{m=1}^{\infty} \sum_{l=1}^{m} P\left\{\sum_{j=1}^{m-1} \mu_{ik}(n-k)=i\right\} q_m(k).$$

§3.3. Local Limit Theorems

The proof of this lemma follows from the relation

$$H_n(N_1,N_2,s) = \prod_{k=N_1}^{N_2} h_k(f_{n-k}(s)) \tag{3.34}$$

and the equality

$$\prod_{k=1}^{n} a_k - \prod_{k=1}^{n} b_k = \sum_{k=1}^{n} (a_k-b_k) \prod_{i=1}^{k-1} a_i \prod_{j=k+1}^{n} b_j.$$

Lemma 3.6. The following inequalities

$$\sup_{\substack{l\geq 1 \\ k\leq N_2}} |a_l^{(1)}(k,n)| \leq \sum_{j=N_1}^{N_2} \alpha(j) \sup_l |Q_1(n-j)|, \tag{3.35}$$

$$\sum_{i=1}^{\infty} i a_i^{(2)}(k,n) \leq \beta(k), \quad n \geq N_2, \quad k \leq n, \tag{3.36}$$

are true for coefficients of the generating functions $A_j(k,n,s)$.

The proof of this lemma follows from definitions of $A_j(k,n,s)$ by using properties of the norma L_1.

Now we go over to the proof of our theorems. Let $L_1(n)$, $L_2(n)$ be positive integer-valued functions such that

$$L_i(n) \to \infty, \quad L_1(n) = o(n), \quad n \to \infty, \quad \limsup_{n \to \infty} L_2(n) n^{-1} < 1.$$

Define the "partial" processes by the relations

$$Z_1(n) = Z(0, L_1(n), n), \quad Z_2(n) = Z(L_1(n), n - L_2(n), n),$$

$$Z_3(n) = Z(n - L_2(n), n, n), \quad H_n^{(i)}(s) = MS^{z_i(n)},$$

$$P_{ij}^{(k)}(n) = P\{Z_k(n) = j | Z_k(0) = i\}.$$

First, using lemmas 3.2, 3.5 and 3.6, we shall prove local limit theorems for the processes $Z_i(n)$. Then we shall deduce the statements of theorems 3.2 and 3.3 from these theorems, choosing suitable functions $L_i(n)$, $i=1, 2$.

Proposition 3.1. If conditions (3.1) and (3.3) are satisfied and $n \to \infty$, $j \to \infty$ such that j/n is bounded, then

$$(\sum_{k=0}^{L_1(n)} \alpha(k))^{-1} (Bn/2)^2 e^{2j/Bn} P_{oj}^{(1)}(n) \to 1.$$

Proof. Putting in Lemma 3.5 $N_1=0$, $N_2=L_1(n)$, we obtain the equality

$$P_{oj}^{(1)}(n) = \sum_{k=0}^{L_1(n)} \alpha(k) \sum_{k=0}^{j} a_{j-i}^{(1)}(k,n) Q_i(n-k) -$$

$$- \sum_{k=o}^{L_1(n)} \sum_{i=1}^{\infty} i(1-f_{n-k}^{i-1}(0))q_i(k) \sum_{i=1}^{j} a_{j-i}^{(1)}(k,n)Q_i(n-k) +$$

$$+ \sum_{k=o}^{L_1(n)} \sum_{l=1}^{j} \sum_{i=1}^{1} a_{l-i}^{(1)}(k,n)Q_i(n-k)a_{j-l}^{(2)}(k,n) = I_1+I_2+I_3. \quad (3.37)$$

Since $\sum_{k=o}^{L_1(n)} \alpha(k)(1-f_{n-k}^{i-1}(0))=o(1)$, $A_1(k,n,s) \to 1$ as $n \to \infty$ uniformly on $s \in D$ and $0 \le k \le L_1(n)$, using the local limit theorem for $\mu(n)$ (Kesten et. al., 1966) we can see that

$$(\sum_{k=o}^{L_1(n)} \alpha(k))^{-1} (Bn/2)^2 e^{2j/Bn} \sum_{k=o}^{L_1(n)} \alpha(n) a_o^{(1)}(k,n)Q_j(n-k)$$

tends to 1 as $n,j \to \infty$ and $j/n \le C < \infty$.

Granting that (Kesten et. al., 1966)

$$\sup_{j,n} n^2 Q_j(n) \le M_1 < \infty, \quad (3.38)$$

under condition (3.1), we obtain the inequality

$$n^2 \sum_{k=o}^{L_1(n)} \alpha(k) \sum_{i=1}^{j-1} a_{j-i}^{(1)}(k,n)Q_i(n-k) \le$$

$$\le 2c_1(1-H_n^{(1)}(0)) \sum_{k=o}^{L_1(n)} \alpha(k) .$$

Therefore, in view of the fact that $H_n^{(1)}(0) \to 1$ as $n \to \infty$, it follows that

$$(\sum_{k=o}^{L_1(n)} \alpha(k))^{-1} n^2 \sum_{k=o}^{L_1(n)} \alpha(k) \sum_{i=1}^{j-1} a_{j-i}^{(1)}(k,n)Q_i(n-k)$$

tends to zero as $n, j \to \infty$ and $j/n \le C < \infty$.

Since $A_1(k,n,1) \le 1$, for $0 \le k \le n$, using (3.38) once again, we can see that $n^2 I_2$ for any sufficiently large n is not greater than $c_2 n^{-1} \sum_{k=0}^{L_1(n)} \beta(k)$, which tends to zero as $n \to \infty$.

Let us consider I_3. It is obvious that

§3.3. Local Limit Theorems

$$I_3 \leq \sum_{k=0}^{L_1(n)} \sup_{1 \leq l \leq j-1} \sum_{i=1}^{l} \alpha(k) a_{l-1}^{(1)}(k,n) Q_i(n-k) \sum_{l=1}^{j-1} a_{j-i}^{(2)}(k,n).$$

By virtue of the estimation of $a_i^{(2)}(k,n)$ from Lemma 3.5, we have for $j \geq 1$:

$$\sum_{l=1}^{j-1} a_{j-i}^{(2)}(k,n) \leq \sum_{l=1}^{\infty} \sum_{j=1}^{\infty} \sum_{t=1}^{j-t} P\left\{\sum_{m=1}^{j-t} \mu_{mk}(n-k)=1\right\} q_j(k) \leq$$

$$\leq (1-f_{n-k}(0))\beta(k). \tag{3.39}$$

On the other hand for some c_3 and for any sufficiently large n

$$n^2 \sup_{1 \leq l \leq j-1} \sum_{i=1}^{l} \alpha(k) a_{l-1}^{(1)}(k,n) Q_i(n-k) \leq c_3 \sum_{i=0}^{j-2} a_i^{(1)}(k,n) \leq c_4 < \infty.$$

It follows from these estimations that $n^2 I_3$ tends to zero as $n, j \to \infty$ and $j/n \leq C < \infty$. Consequently, by virtue of (4.7), the $P_{oj}^{(1)}(n)$ multiplied by the quantity in the statement of the theorem tends to 1 as $n, j \to \infty$ and $j/n \leq C$. The proposition is proved.

Proposition 3.2. Let conditions (3.1), (3.2) be satisfied and $\beta(n)=o(\alpha(n))$. If $n, j \to \infty$ such that $j/L_2(n)$ is bounded, then

$$\frac{Bj}{2\alpha(n)} \left(1 - \frac{2}{BL_2(n)}\right)^{-j} P_{oj}^{(3)}(n) \to 1.$$

Proof. Putting $N_1 = n - L_2(n)$, $N_2 = n$ in Lemma 3.5, we have:

$$P_{oj}^{(3)}(n) = \sum_{k=n-L_2(n)}^{n} \sum_{i=1}^{j} a_{j-i}^{(1)}(k,n) Q_i(n-k) -$$

$$- \sum_{k=n-L_2(n)}^{n} \sum_{i=1}^{\infty} i(1-f_{n-k}^i(0)) q_i(k) \sum_{i=1}^{j} a_{j-i}^{(1)}(k,n) Q_i(n-k) +$$

$$+ \sum_{k=n-L_2(n)}^{n} \sum_{l=1}^{j-1} \sum_{i=1}^{l} a_{l-i}^{(1)}(k,n) Q_i(n-k) a_{j-l}^{(2)}(k,n) = I_1 + I_2 + I_3. \tag{3.40}$$

Since $\alpha(n)$ is a regularly varying function and $a_o^{(1)}(k,n) \to 1$ as $n \to \infty$ uniformly on $k \in [n-L_2(n), n]$, using Lemma 3.6, we obtain that the sum

$$\frac{Bj}{2\alpha(n)}\left(1-\frac{2}{BL_2(n)}\right)^{-j}\sum_{k=n-L_2(n)}^{n}\alpha(k)a_o^{(1)}(k,n)Q_j(n-k)$$

tends to 1 as $n, j \to \infty$ and $j/n \leq C < \infty$.

Let $0 < \varepsilon < 1$. Consider the sum

$$\sum_{k=n-L_2(n)}^{n}\sum_{i=1}^{j-1}a_{j-i}^{(1)}(k,n)Q_i(n-k) = \sum_{k=n-L_2(n)}^{n}\sum_{i=1}^{[j\varepsilon]}a_{j-i}^{(1)}(k,n)Q_i(n-k) +$$

$$+ \sum_{k=n-L_2(n)}^{n}\sum_{i=[j\varepsilon]}^{j-1}a_{j-i}^{(1)}(k,n)Q_i(n-k) = R_1 + R_2. \qquad (3.41)$$

If we use (3.35), then we obtain the following estimation for the first term:

$$R_1 \leq M_2\alpha(n)j^{-1}\sup_1 1G(1,1)\sum_{i=1}^{[\varepsilon j]}G(1,i).$$

Taking into account (3.4) and the fact that (Kesten et. al., 1966, p. 582)

$$\sum_{i=1}^{j}G(1,i) \sim \frac{2\ln j}{B}, \quad j \to \infty,$$

we have, for some M_3 and for any sufficiently large n and j,

$$R_1 \leq M_3\alpha(n)j^{-1}\ln j. \qquad (3.42)$$

Using relation (3.4) again, we obtain the estimate

$$R_2 \leq M_4 j^{-1}\sup_{n-L_2(n)\leq k\leq n}\sum_{i=1}^{\infty}a_i^{(1)}(k,n).$$

Therefore, by virtue of (3.41) and (3.42) I_1 multiplied by $(Bj/2\alpha(n))(1-2/BL_2(n))^{-j}$ tends to 1 as $n, j \to \infty$ and $j/L_2(n) \leq C < \infty$.

Similarly we obtain that for some M_5 and for a sufficiently large n and j

$$I_2 \leq M_5 j^{-1}\sup_{n-L_2(n)\leq k\leq n}\beta(k). \qquad (3.43)$$

Let us estimate I_3. Consider the sum

$$R_3 = \sum_{k=n-L_2(n)}^{n}\sum_{l=1}^{[j\varepsilon]}\sum_{i=1}^{l}a_{l-i}^{(1)}(k,n)Q_i(n-k)a_{j-1}^{(2)}(k,n).$$

Since $a_i^{(1)}(k,n)$ is non-negative, using (4.8) we have:

§3.3. Local Limit Theorems

$$R_3 \leq M_6 \sum_{k=n-L_2(n)}^{n} (n-k)^{-2} \sum_{i=0}^{[j\varepsilon]} a_i^{(1)}(k,n) \sum_{l=1}^{[j\varepsilon]} a_{j-1}^{(2)}(k,n).$$

Granting (3.36), we obtain finally

$$R_3 \leq \frac{M_6}{j(1-\varepsilon)} \sum_{k=n-L_2(n)}^{n-1} \beta(k)(n-k)^{-2} \leq M_7 j^{-1} \sup_{n-L_2(n) \leq k \leq n} \beta(k). \tag{3.44}$$

Using relation (3.39) as in the estimation of I we can that

$$R_4 \leq M_8 j^{-1} \sup_{n-L_2(n) \leq k \leq n} \beta(k). \tag{3.45}$$

Since $I_3 \leq R_3 + R_4$, it follows from (3.43)-(3.45) that $j(I_2+I_3)/\alpha(n)$ tends to zero as $n, j \to \infty$ and $j/L_2(n) \leq C < \infty$. The proposition is proved.

Proposition 3.3. Let the conditions of Proposition 3.2 be satisfied. If $n, j \to \infty$ such that $j/L_2(n)$ is bounded, then

$$P_{oj}^{(2)}(n) = 0\left(\frac{1}{n^2} \sum_{k=L_1(n)}^{n} \alpha(k) + \alpha(n) \sum_{k=L_2(n)}^{n} Q_j(k)\right) + o\left(\frac{\alpha(n)}{j}\right). \tag{3.46}$$

Proof. It follows from Lemma 3.5 that

$$0 \leq P_{oj}^{(2)}(n) \leq I_1 + I_2, \tag{3.47}$$

$$I_1 = \sum_{k=L_1(n)}^{n-L_2(n)} \alpha(k) \sum_{i=1}^{j} a_{j-i}^{(1)}(k,n) Q_i(n-k),$$

$$I_2 = \sum_{k=L_1(n)}^{n-L_2(n)} \alpha(k) \sum_{l=1}^{j-1} \sum_{i=1}^{l} a_{l-i}^{(1)}(k,n) Q_i(n-k) a_{j-l}^{(2)}(k,n).$$

Let $0 < \varepsilon < 1$ such that the $[n\varepsilon]$ is not greater than $n-L_2(n)$ for any sufficiently large n. We shall divide I_2 in two parts: according to k such that $L_1(n) \leq k \leq [n\varepsilon]$ and $[n\varepsilon] < k \leq n-L_2(n)$. Using (3.38) and the boundedness of $\sum_{i=0}^{\infty} a_i^{(1)}(k,n)$, we obtain that the first part of the sum is less than $M_9 n^{-2} \sum_{k=L_1(n)}^{n} \alpha(k)$. It is obvious that

$$\sum_{k=[n\varepsilon]+1}^{n-L_2(n)} \alpha(k) a_o^{(1)}(k,n) Q_j(n-k) \leq M_{10} \sum_{k=L_2(n)}^{n} Q_j(k).$$

Since the estimations of R_1 and R_2 from (3.41) are true with corresponding changes,

$$\sum_{k=[n\varepsilon]+1}^{n-L_2(n)} \alpha(k) \sum_{i=1}^{j-1} a_{j-i}^{(1)}(k,n)Q_i(n-k) = o\left(\frac{\alpha(n)}{j}\right)$$

as $n \to \infty$. Therefore

$$I_1 = O\left(n^{-2} \sum_{k=L_1(n)}^{n} \alpha(k) + \alpha(n) \sum_{k=L_2(n)}^{n} Q_j(k)\right) + o\left(\frac{\alpha(n)}{j}\right).$$

If we divide I_2 into two parts in exactly the same way as I_1, then using (3.38), (3.39), we can see that

$$\sum_{k=L_1(n)}^{[n\varepsilon]} \sum_{l=1}^{j-1} \sum_{i=1}^{l} a_{l-i}^{(1)}(k,n)Q_i(n-k)a_{j-l}^{(2)}(k,n) = o(n^{-2}).$$

In order to estimate the remaining parts of I_2, we shall divide the sum according to l into two parts: according to the values of l such that $1 \le l \le [j\varepsilon]$ and $[j\varepsilon] < l \le j-1$. We can estimate the first part using relation (3.36) and the estimates of R_1 and R_2 from (3.41). The second part can be estimated using inequality (3.39) and the estimates of R_1 and R_2. Thus, as $n, j \to \infty$,

$$I_2 = o(n^{-2} + \alpha(n)j^{-1}). \tag{3.49}$$

Proposition 3.3 follows from (3.47)-(3.49).

Before going over the proof of the basic theorems, we will mention two subsidiary statements.

Proposition 3.4. If the conditions of Proposition 3.2 are satisfied and $n, j \to \infty$, such that $j/L_2(n) \le C < \infty$, then

$$P_{oj}(n) \sim P_{oj}^{(1)}(n) + P_{oj}^{(2)}(n) + P_{oj}^{(3)}(n).$$

The proof follows from the equality

$$Z(n) = Z_1(n) + Z_2(n) + Z_3(n)$$

and from the independence of processes $Z_i(n)$.

Proposition 3.5. Let $Z^*(n)$ be a Galton-Watson process with the same distribution of the number of direct offspring as $Z(n)$ and with a number of immigrating particles $\{\eta_n, n \in N_0\}$, and, in addition, let function $h_n^*(s) = Ms^{\eta_n}$ be defined by relations

$$h_o^*(s) = s^i h_o(s), \quad h_k^*(s) = h_k(s), \quad k \ge 1.$$

§3.3. Local Limit Theorems   153

Then $P_{ij}(n) = P\{Z^*(n)=j \mid Z^*(0)=0\}$.

This proposition follows from the definition of the process.

Proof of Theorem 3.2. Note that if $j, n \to \infty$, such that $\ln j/\ln n \in K(0,1)$, then $j \ln n/n \to 0$. Choosing function $L_2(n)$ such that $\ln L_2(n) \sim \ln n$, $n \to \infty$, we have $j/L_2(n) \to 0$. Then, it follows from Lemma 3.2 that

$$\sum_{k=L_2(n)}^{n} Q_j(k) = o(j^{-1}), \quad n, j \to \infty.$$

Let $\alpha(n) = n^{-\alpha}L(n)$, $\alpha \geq 0$, where $L(n)$ is a slowly varying function at infinity. It follows from the condition $\liminf_{n \to \infty} \theta_1(n) > 0$ that $0 \leq \alpha < 1$. If $0 \leq \alpha < 1$, then $j \sum_{k=0}^{L_1(n)} \alpha(k)/n^2 \alpha(n) \to 0$ as $n, j \to \infty$. If $\alpha = 1$, then we can choose the function $L_1(n)$ such that

$$\sum_{k=0}^{L_1(n)} \alpha(k) \sim \sum_{k=0}^{n} \alpha(k).$$

Since

$$jn^{-2} \sum_{k=0}^{n} \alpha(k)/\alpha(n) = jn^{-1}\ln n/\theta_1(n)$$

tends to zero as $n, j \to \infty$, we obtain from propositions 3.1-3.4 that as $n, j \to \infty$, $\ln j/\ln n \in K(0,1)$:

$$P_{oj}(n) \sim 2\alpha(n)/Bj \qquad (3.50)$$

It follows from Proposition 3.5 that in (3.50) the probability $P_{oj}(n)$ can be replaced by $P_{ij}(n)$, where i is fixed. Theorem 3.2 is thus proved.

Proof of Theorem 3.3. It follows from the condition $\limsup_{n \to \infty} \theta_1(n) < \infty$ that $\alpha \geq 1$. Choose $L_i(n)$, $i=1, 2$, such that

$$\sum_{k=0}^{L_1(n)} \alpha(k) \sim \sum_{k=0}^{n} \alpha(k), \quad L_2(n) \sim [n\varepsilon], \quad 0 < \varepsilon < 1.$$

Then we have that, if $j/n \in K(0,C)$, then $j/L_2(n) \in K(0, \varepsilon C)$. Since

$$\sum_{k=L_1(n)}^{n} \alpha(k) = o\left(\sum_{k=0}^{n} \alpha(k)\right), \quad n^2\alpha(n)/j \sum_{k=0}^{n} \alpha(k) = n\theta_1(n)/j\ln n \to 0$$

as $n, j \to \infty$, we conclude from propositions 3.1-3.4 that

$$P_{oj}(n) \sim \left(\frac{2}{Bn}\right)^2 e^{-2j/Bn} \sum_{k=0}^{n} \alpha(k).$$

Using Proposition 3.5, we obtain that

$$\left(\sum_{k=0}^{n} \alpha(k)+i\right)^{-1} \left(\frac{Bn}{2}\right)^2 e^{2j/Bn} P_{ij}(n)$$

tends to 1 as $n, j \to \infty$ and $j/n \in K(0,C)$. Theorem 3.3 is proved.

§3b. Occupation of a Fixed State

It is known that the reflexivity conditions of Markov chains are defined by the asymptotics of the hit probability to a fixed state. From this point of view it is interesting to investigate the asymptotical behaviour of $P_{ij}(n)$ when $n \to \infty$ and remaining parameters are fixed.

It has been proved by Pakes (1975) that, if immigration is homogeneous, under some conditions for critical processes there exist the limits

$$0 \leq \nu_j = \lim_{n \to \infty} n^\gamma P_{ij}(n), \quad \gamma \in (0, \infty), \quad i, j \in N.$$

It has been found that in the case of decreasing immigration the structure of the asymptotics of these probabilities is more extended, and essentially depends on the rate of decrease of immigration, on a stationary measure and on a Green's function of the process without immigration. In addition it is found that the hit probability to the state zero is "asymptotically greater" than hit probabilities to other states (see Rahimov, 1986a).

Let $\{\nu_i, i \in N\}$ be a stationary measure, $G(i,j) = \sum_{n=0}^{\infty} Q_{ij}(n)$, $i \in N_o$, be a Green's function of $\mu(n)$ and let $\alpha(n)$ be a slowly varying function such that

$$\alpha(n) \sum_{k=1}^{n} \left| \left(\frac{Bk}{2} - \frac{1}{1-s}\right)^{-1} -1 + f_k(s)\right| = o(1), \quad 0 \leq s < 1. \tag{3.51}$$

Condition (3.51) is fulfilled, for example, when (3.1) is satisfied and $\alpha(n) \to 0$, $n \to \infty$, (see Zubkov, 1972).

Theorem 3.4. Let conditions (3.2), (3.51) be satisfied and let $\beta(n) \to 0$ as $n \to \infty$. Then

$$P_{oi}(n) \sim n^{-\gamma_n}, \quad P_{ij}(n) = o(n^{-\gamma_n}).$$

§3.3. Local Limit Theorems

Theorem 3.5. Let conditions (3.2) and (3.51) be satisfied, $i \in N_o$, $j \in N$ and let $\beta(n) = O(\alpha(n) + n^{-2} \sum_{k=0}^{n} \alpha(k))$. Then,

$1°$ if $\alpha(n)n^2 \to \infty$, then $P_{ij}^{(n)} \sim n^{-\gamma_n} \alpha(n) G(1, j)$;

$2°$ if $\alpha(n)n^2 \to 0$, then $P_{ij}(n) \sim 2B^{-1} n^{-2} \nu_j (i + \sum_{k=0}^{\infty} \alpha(k))$;

$3°$ if $\alpha(n)n^2 \to C \in (0, \infty)$, then

$$P_{ij}(n) \sim \frac{2}{Bn^2} \left[\frac{BC}{2} G(1, j) + \nu_j \left(\sum_{k=0}^{\infty} \alpha(k) + i \right) \right].$$

CHAPTER IV.

THE ASYMPTOTIC BEHAVIOR OF FAMILIES OF PARTICLES IN THE BRANCHING PROCESS

The results of Chapter II indicate that it is possible to investigate different models of branching processes with immigration using methods of the theory of summation of a random number of random variables. In this chapter we will use these methods to study some new characteristics of branching processes without immigration.

We will consider a scheme of evolution and reproduction of some particles. These particles live independently of each other in a random time L and generate a random number ν of new particles. These new particles undergo analogous transformation. Let us assume that there is an ancestor at the time t=0. We will use the same notations as in §2.2, namely, denote by (1) the initial particle and by $\alpha' = (i_1, \ldots, i_k, j)$ the jth direct offspring of $\alpha = =(i_1, \ldots, i_k)$. Then the set $\mathcal{A}_t \in E$, where E is the space of all finite subsets of

$$\bigcup_{k=1}^{\infty} N^k, \quad N^k = N^{k-1} N, \quad N^1 = N = \{1, 2, \ldots\},$$

corresponds to the population at time t.

The main object of investigation of the theory of branching processes is the variable $\mu(t) = |\mathcal{A}_t|$; $|\mathcal{A}|$ is the power (the number of elements) of the set $\mathcal{A}$. In addition the relation

$$\mu(t) = \sum_{\alpha \in \mathcal{A}_\tau} \mu^{(\alpha)}(t-\tau), \quad t > \tau, \qquad (0.1)$$

where $\mu^{(\alpha)}(t)$ is the branching process generated by particle α, is the starting point of the investigation. Since the processes $\mu^{(\alpha)}(t)$ are independent and identically distributed, the study of $\mu(t)$ can be reduced to the analysis of a corresponding equation for generating functions. However many characteristics of the population $\mathcal{A}_t$ are not representable in the form (0.1) and, therefore, it is not possible to study them using the traditional methods of the theory of branching processes.

§4.1. Sums of Dependent Indicators

Let $\mathcal{B}(\tau,t)=\{\alpha\in\mathcal{A}_\tau: \mu^{(\alpha)}(t-\tau)\neq 0\}$ be the set of particles in $\mathcal{A}_\tau$ having a nonempty set of offspring at time t. We shall define the random variables (the "distance") $\rho(\alpha,\beta)$, $\alpha,\beta \in (\tau,t)$, by the relation:

$$\rho(\alpha,\beta) = |\mu^{(\alpha)}(t-\tau) - \mu^{(\beta)}(t-\tau)|.$$

Let $d\in N_0=\{0,1,\ldots\}$ and let $\chi(A)$ be an indicator function of an event A. For arbitrary moments of time τ and t, with $\tau \leq t$, let

$$z_r(\tau,t,d) = \frac{1}{r!} \Sigma \chi(\alpha_{i_1},\ldots,\alpha_{i_r}), \qquad (0.2)$$

where summation with respect to $\alpha_{i_1},\ldots,\alpha_{i_r} \in \mathcal{B}(\tau,t)$ and

$$\chi(\alpha_1,\ldots,\alpha_r) = \prod_{1\leq i<j\leq r} \chi\{\alpha_i\neq\alpha_j\}\ \chi\{\rho(\alpha_i,\alpha_j)\leq d\}, \quad r\geq 2.$$

Obviously, $z_r(\tau,t,d)$ is equal to the number of r-tuples of particles in $\mathcal{A}_\tau$ that have at time t nonempty offspring sets differing from each other by at most d. In particular, $z_2(\tau,t,0)$ is the number of pairs of particles in $\mathcal{B}(\tau,t)$ having the same number of offspring at time t.

The variable $z_2(\tau,t,d)$ can be connected with a scheme of allocation of some particles into cells (Kolchin et.al. (1978), see §4.2).

We shall investigate the asymptotic properties of the process $z_r(\tau,t,d)$ and of some other characteristics of the population $\mathcal{A}_t$ when $\tau,t\to\infty$, r=2 and $\mu(t)$ is a Galton-Watson process. This investigation will be realized in sections 4.2 and 4.3. In §4.1 we will prove some limit theorems for random sums of dependent indicators which will be the subsidiary results for the next sections. The methods used here are applicable for the study of characteristics of more general models of branching processes and for r>2.

§4.1. SUMS OF DEPENDENT INDICATORS

§1a. Sums of Functions of Independent Random Variables

Let $\{\xi_{in}, i\in N\}$ be a sequence of independent and identically distributed random variables for each $n\in N$ and let $\{\nu_n, n\in N\}$ be a sequence of random variables with values in the set N_0. Assume that variables ν_n and $\{\xi_{in}, i\in N\}$ are independent for each $n\in N$ and put

IV. Families of Particles in Branching Processes

$$T_n^{(r)}(\nu_n) = \begin{cases} \sum_{1 \le i_1 < \ldots < i_r \le \nu_n} g_n(\xi_{i_1 n}, \ldots, \xi_{i_r n}), & \nu_n \ge r, \\ 0, & \nu_n < r, \end{cases}$$

where $g_n(\cdot)$ are symmetric functions taking the values 0 and 1.

We also consider the sum

$$S_{\nu_n} = \eta_1 + \eta_2 + \ldots + \eta_{\nu_n},$$

where $\eta_1, \eta_2, \ldots$ is a scheme of series of, generally speaking, arbitrarily dependent random variables taking the values 0 and 1.

Sums of dependent indicators with a determined number of terms have been studied by many authors. So limit theorems for S_n have been proved by Sevast'yanov (1972) and Ambrosimov (1976), who obtained some general conditions for convergence of the distribution of this sum to the Poisson and Normal distributions. Estimation of the variation distance between distribution of $T_n^{(r)}(n)$ and approximating distributions have been studied by Zubkov (1977). The papers of Silverman and Brown (1978) and Barbour and Eagleson (1983) have also contributed to limit theorems for $T_n^{(r)}(n)$.

Let us introduce the following conditions:

Condition 1.1. As $n \to \infty$

$$C_{\nu_n}^r M\left[g_n(\xi_{1n}, \ldots, \xi_{rn})\right] \xrightarrow{D} \nu, \quad P\{\nu \le x\} = F(x),$$

where, as before, D means the convergence in distribution.

Condition 1.2. The function $g_n(\cdot)$ is such that if

$$\lim_{n \to \infty} \binom{l_n}{r} M\left[g_n(\xi_{1n}, \ldots, \xi_{rn})\right] = \lambda > 0$$

for a sequence $l_n \to \infty$, $n \to \infty$, then

$$\lim_{n \to \infty} l_n^{2r-1} M\left[g_n(\xi_{1n}, \ldots, \xi_{rn}) g_n(\xi_{1n}, \ldots, \xi_{r-1 n}, \xi_{r+1 n})\right] = 0.$$

Theorem 1.1. Under conditions 1.1 and 1.2

$$\lim_{n \to \infty} P\{T_n^{(r)}(\nu_n) = k\} = \frac{1}{k!} \int_0^\infty x^k e^{-x} dF(x). \quad (1.1)$$

Note that Theorem 1.1 was proved by Silverman and Brown (1978) in the case of a determined number of terms. Now we represent an estimation of the rate of convergence in Theorem 1.1. Following Zubkov (1977), we introduce some notations. Denote by $I^{(1)}$ and $I_k^{(1)}$ ordered r-tuples $(i_1, \ldots, i_r)$ and $(i_{1k}, \ldots, i_{rk})$ of positive integers from 1 to l, and denote by $\{I^{(1)}\}$ the set

§4.1. Sums of Dependent Indicators

$\{i_1, i_2, \ldots, i_r\}$. Let

$$Y(r,1) = \{I^{(1)} = (i_1, \ldots, i_r): 1 \leq i_1 < \ldots < i_r \leq 1\},$$

$$\theta^{(n)}_{I^{(1)}} = g_n(\xi_{i_1 n}, \ldots, \xi_{i_r n}),$$

$$b_{I^{(1)}_1 \ldots I^{(1)}_k} = P\left\{\theta^{(n)}_{I^{(1)}_1} = \ldots = \theta^{(n)}_{I^{(1)}_k} = 1\right\}$$

if $I^{(1)}_1, \ldots, I^{(1)}_k \in Y(r,1)$ are mutually different, and

$$b_{I^{(1)}_1 \ldots I^{(1)}_k} = 0$$

otherwise;

$$\lambda(r,1) = \sum_{I^{(1)} \in Y(r,1)} b_{I^{(1)}}, \quad Y^k(r,1) = \{(I^{(1)}_1, \ldots, I^{(1)}_k): I^{(1)}_j \in Y(r,1)\},$$

$$Y^k_o(r,1) = \{(I^{(1)}_1, \ldots, I^{(1)}_k) \in Y^k(r,1): \{I^{(1)}_i\} \cap \{I^{(1)}_j\} = \emptyset, \; (i \neq j)\},$$

$$R(I^{(1)}_1, \ldots, I^{(1)}_k) \in \{I^{(1)} \in Y(r,1): \{I^{(1)}\} \cap \{I^{(1)}_j\} = \emptyset, \; j=1,\ldots,k\},$$

$$S^{(1)}_k = \sum_{Y^k_o(r,1)} \sum_{I^{(1)} \in R(I^{(1)}_1, \ldots, I^{(1)}_k)} b_{I^{(1)}_1 \ldots I^{(1)}_k, I^{(1)}},$$

$$T^{(1)}_k = \sum_{Y^k_o(r,1)} \sum_{I^{(1)} \in R(I^{(1)}_1, \ldots, I^{(1)}_k)} b_{I^{(1)}_1} b_{I^{(1)}_2} \cdots b_{I^{(1)}_k} b_{I^{(1)}}.$$

Zubkov (1977) provides one estimation which will be expressed by

$$\rho_r(n,1) = \left[T^{(1)}_o + 2D(m,1)(S_1+T_1)\right] / \left[1-T^{(1)}_o - C(m,1)S_1\right]_+,$$

where

$$S_1 = \sum_{k=1}^{r} \frac{S^{(1)}_k}{k!}, \quad T_1 = \sum_{k=1}^{r} \frac{T^{(1)}_k}{k!}, \quad D(m,1) = C(m,1) + \alpha_m(r,1)e^{\lambda(r,1)},$$

IV. Families of Particles in Branching Processes

$$\alpha_m(r,1) = a_1^{(1)} r(m+r+\lambda(r,1)),$$

$$a_1^{(1)} = \max_{1 \le i \le 1} \sum_{I^{(1)} \in Y(r,1): i \in I^{(1)}} b_{I^{(1)}},$$

$$C(m,1) = 1 + \alpha_m(r,1) + \alpha_m^2(r,1) + \ldots + \alpha_m^{m-1}(r,1).$$

We also put

$$F_n(x) = P\{(\binom{r}{\nu_n}) M[g_n(\xi_{1n}, \ldots, \xi_{rn})] \le x\},$$

$$\delta(n) = \sup_x |F_n(x) - F(x)|,$$

$$\Delta_n(k) = |P\{T_n^{(r)}(\nu_n) = k\} - \frac{1}{k!} \int_0^\infty x^k e^{-x} dF(x)|.$$

Theorem 1.2. The following inequality

$$\sup_k \Delta_n(k) \le M \rho_r(n, \nu_n) + 2\delta(n), \qquad (1.2)$$

is valid and, if

$$0 < \beta_n \equiv M\nu_n < \infty, \quad \gamma_n \equiv \max_{0.5\beta_n \le 1 \le 1.5\beta_n} \rho_r(n,1) + 4 \frac{M|\nu_n - \beta_n|}{\beta_n},$$

then

$$\sup_k \Delta_n(k) \le \gamma_n + 2\delta(n). \qquad (1.3)$$

Proof of Theorem 1.1. Let

$$A_n = Mg_n(\xi_{1n}, \ldots, \xi_{rn}).$$

It is not difficult to see that for any $q > 0$

$$P\{T_n^{(r)}(\nu_n) \le k\} = \sum_{l_n=0}^\infty P\{T_n^{(r)}(\nu_n) \le k\} P\{\nu_n = l_n\} =$$

$$= \sum_{i=0}^\infty \sum_{i \le \binom{r}{1} A_n q < i+1} P\{T_n^{(r)}(\nu_n) \le k\} P\{\nu_n = l_n\}.$$

Let $p > 0$ be such that pq is an integer. Then

$$P\{T_n^{(r)}(\nu_n) \le k\} = \sum_{i=0}^{pq-1} \sum_{i \le \binom{r}{1} A_n q < i+1} P\{T_n^{(r)}(1_n) \le k\} P\{\nu_n = l_n\} +$$

§4.1. Sums of Dependent Indicators

$$+ \sum_{i=pq} \sum_{\substack{i \leq \binom{r}{1}A_n q < i+1}} P\{T_n^{(r)}(1_n) \leq k\} P\{\nu_n = 1_n\} = I_1 + I_2. \qquad (1.4)$$

We use the following version of Silverman-Brown's theorem.

Theorem A. Let $g_n(\cdot)$ be a family of symmetric functions of m arguments taking values 0 and 1. Suppose that for some λ and for some sequence of integers $l_n \to \infty$, $n \to \infty$,

$$\binom{m}{1_n} M[g_n(\xi_{1n}, \ldots, \xi_{mn})] \to \lambda,$$

$$l_n^{2m-1} M[g_n(\xi_{1n}, \ldots, \xi_{mn}) g_n(\xi_{1n}, \ldots, \xi_{m-1n}, \xi_{m+1n})] \to 0.$$

Then as $n \to \infty$

$$\sum_{1 \leq i_1 \leq \ldots \leq i_m \leq l_n} g_n(\xi_{i_1 n}, \ldots, \xi_{i_m n}) \xrightarrow{d} \xi,$$

where the distribution of ξ is a Poisson with parameter λ.

If we denote by $\nu(a)$ a random variable with Poisson distribution of the parameter a, then

$$\pi(a, k) = P\{\nu(a) \leq k\}$$

is a decreasing function of a for any fixed k. It follows from Theorem A and from assumptions on functions $g_n(\cdot)$ that, if $l_n \to \infty$ as $n \to \infty$ such that $i \leq \binom{r}{1}A_n q < i+1$, then for $i = 0, 1, \ldots, pq-1$

$$\pi\left(\frac{i+1}{q}, k\right) \leq \liminf_{n \to \infty} P\{T_n^{(r)}(1_n) \leq k\} \leq$$

$$\leq \liminf_{n \to \infty} P\{T_n^{(r)}(1_n) \leq k\} \leq \pi(i/q, k). \qquad (1.5)$$

Using relations (1.5) and granting Condition 1.1, we have for any p and q:

$$\limsup_{n \to \infty} I_1 \leq \sum_{i=0}^{pq-1} \pi(1/q, k) \left[F\left(\frac{i+1}{q}\right) - F\left(\frac{i}{q}\right)\right],$$

$$\limsup_{n \to \infty} I_1 \geq \sum_{i=0}^{pq-1} \pi(i+1/q, k) \left[F\left(\frac{i+1}{q}\right) - F\left(\frac{i}{q}\right)\right].$$

Since sums on the right-hand side of the last relations tend to the same integral for any p and $q \to \infty$,

$$\lim_{n\to\infty} I_1 = \int_0^p \pi(x,k) dF(x). \tag{1.6}$$

The second term in (1.4) will be arbitrarily small for sufficiently large p. Therefore, by virtue of the arbitrariness of p, we obtain the following relation

$$P\{T_n^{(r)}(1_n) \le k\} = \int_0^p \pi(x,k) dF(x) + o(1), n\to\infty,$$

which is equivalent to the statement of Theorem 1.1.

Proof of Theorem 1.2. Let us prove (1.2). Consider the relation

$$\Delta_n(k) \le I_1(n,k) + I_2(n,k) + I_3(n,k), \tag{1.7}$$

where

$$I_1(n,k) = \sum_{1_n=0}^{\infty} \left| P\{T_n^{(r)}(1_n) \le k\} - \frac{\lambda^k(r,1_n)}{k!} e^{-\lambda(r,1_n)} \right| P\{\nu_n = 1_n\},$$

$$I_2(n,k) = \left| \sum_{1_n=0}^{\infty} \frac{\lambda^k(r,1_n)}{k!} e^{-\lambda(r,1_n)} P\{\nu_n=1_n\} - \int_0^\infty \frac{x^k}{k!} e^{-x} dF_n(x) \right|,$$

$$I_3(n,k) = \left| \int_0^\infty \frac{x^k}{k!} e^{-x} d(F_n(x) - F(x)) \right|.$$

Using Theorem 2 of Zubkov (1977), we obtain:

$$\sup_k I_1(n,k) \le \sum_{k=0}^{\infty} I_1(k,n) \le M\rho_r(n,\nu_n). \tag{1.8}$$

It follows from the definition of $F_n(x)$ that, in our notations, $F_n(x) = P\{\lambda(r,\nu_n) \le x\}$ and, therefore, the following equality

$$\int_0^\infty \frac{x^k}{k!} e^{-x} dF_n(x) = \sum_{1_n=0}^{\infty} \frac{\lambda^k(r,1_n)}{k!} e^{-\lambda(r,1_n)} P\{\nu_n=1_n\}$$

is true. This shows that $I_2(n,k)=0$ for any k and n.

Finally we have for any k:

$$I_3(n,k) = \left| \int_0^\infty (F_n(x)-F(x)) \frac{k-x}{k!} e^{-x} x^{k-1} dx \right| \le 2\delta(n). \tag{1.9}$$

The proof of inequality (1.2) follows from (1.7)-(1.8).

§4.1. Sums of Dependent Indicators

Now we shall prove (1.3). Consider the quantity $I_1(n,k)$ in relation (1.7). Let $\mathcal{K}_n = \{1_n : |1_n - \beta_n| \leq \beta_n/2\}$, and $\mathcal{K}'_n = \{1_n : |1_n - \beta_n| > \beta_n/2\}$. Then, for any n, $k \in N_0$,

$$I_1(n,k) \leq \sum_{i=0}^{\infty} \sum_{1_n \in \mathcal{K}_n} |P\{T_n^{(r)}(1_n) = i\} - \frac{\lambda^i(r, 1_n)}{i!} e^{-\lambda(r, 1_n)}| P\{\nu_n = 1_n\}$$

$$+ \sum_{i=0}^{\infty} \sum_{1_n \in \mathcal{K}'_n} |P\{T_n^{(r)}(1_n) = i\} - \frac{\lambda^i(r, 1_n)}{i!} e^{-\lambda(r, 1_n)}| P\{\nu_n = 1_n\}$$

$$= S_1 + S_2. \qquad (1.10)$$

Using Theorem 2 of Zubkov (1977) again, we have:

$$S_1 \leq \sum_{1_n \in \mathcal{K}'_n} \rho(n, 1_n) P\{\nu_n = 1_n\} \max_{0.5\beta_n \leq 1 \leq 1.5\beta_n} \rho(n, 1_n). \qquad (1.11)$$

On the other hand by virtue of Chebyshev's inequality

$$S_2 \leq 2 \sum_{1_n \in \mathcal{K}'_n} P\{\nu_n = 1_n\} \leq \frac{M|\nu_n - \beta_n|}{\beta_n}. \qquad (1.12)$$

It follows from relations (1.10)-(1.12) that

$$\sup_k I_1(n,k) \leq \gamma_n. \qquad (1.13)$$

Using the estimates obtained in the proof of the first part of the theorem it is not difficult to see that inequality (1.3) is valid. Theorem 1.2 is now proved.

§1b. Sampling Sums of Dependent Indicators

Now we go over to the results for S_ν. In the scheme of series S_{ν_n} we put

$$b_{i_1 \ldots i_r} = P\{\eta_{i_1} = \eta_{i_2} = \ldots = \eta_{i_r} = 1\},$$

where $1 \leq i_k \leq n$ are mutually different, $k = 1, 2, \ldots, r$,

$$Y(r,n) = \{(i_1, \ldots, i_r) : 1 \leq i_1 < \ldots < i_r \leq n\}, \quad r \geq 2, \; n \geq 2.$$

We denote by $I(r,n) \subset Y(r,n)$ the "exclusive" sets introduced by Sevast'yanov (1972), and put

IV. Families of Particles in Branching Processes

$$\zeta(m) = \sum_{i=1}^{m} b_i, \quad a(j,m,n) = \frac{1}{j!} \zeta^{1-m}(n) e^{\zeta(n)}, \quad \Delta(m,n) = \{m, m+1, \ldots, n\},$$

$\zeta_n^* = [\zeta(l_n)]$, where $l_n \in N_0$, $[x]$ is the integral part of x.

Theorem 1.3. Let the following conditions be satisfied: for a sequence of integers $l_n \to \infty$, $n \to \infty$,

$$\zeta(l_n) \to \infty, \quad \max_{i \in \Delta(1, l_n)} b_i e^{2\zeta(l_n)} \to 0, \quad n \to \infty; \tag{1.14}$$

there are exclusive sets $I(r, l_n)$, $r \in \Delta(2.9\zeta_n^*)$, such that

$$a(j, m, l_n) \sum_{I(m+j, l_n)} [b_{i_1 \ldots i_{m+j}} + b_{i_1} \ldots b_{i_{m+j}}] \to 0 \tag{1.15}$$

as $n \to \infty$ uniformly according to $m \in \Delta(1, 2\zeta_n^*)$ and $j \in \Delta(0.9\zeta_n^* - m)$;

$$\left| \frac{b_{i_1 \ldots i_r}}{b_{i_1} \ldots b_{i_r}} - 1 \right| e^{2\zeta(l_n)} \to 0, \quad n \to \infty, \tag{1.16}$$

uniformly on $(i_1, \ldots, i_r) \in I(r, l_n)$ and on $r \in (1.9\zeta_n^*)$. If the probabilities $b_{i_1 \ldots i_r}$ are such that the fulfillment of these conditions follows from (1.14)-(1.16) for any sequence $l_n' \to \infty$, $n \to \infty$, for which

$$0 < a \leq \frac{\zeta(l_n')}{\zeta(l_n)} \leq b < \infty, \quad n \in N \tag{1.17}$$

and $\zeta(\nu_n)/\zeta(l_n) \xrightarrow{d} \nu$ as $n \to \infty$, then

$$\lim_{n \to \infty} P\left\{ \frac{S_{\nu_n} - \zeta(\nu_n)}{\sqrt{\zeta(l_n)}} < x \right\} = \int_0^\infty \Phi\left(\frac{x}{\sqrt{u}} \right) dT(u),$$

where $T(x)$ is the distribution of ν.

Proof. It follows from the total probability formula that for any $q>0$

$$P\left\{ \frac{S_{\nu_n} - \zeta(\nu_n)}{\sqrt{\zeta(l_n)}} < x \right\} = \sum_{i=0}^{\infty} \sum_{\frac{i}{q} \leq \frac{\zeta(1)}{\zeta(l_n)} < \frac{i+1}{q}} P\left\{ \frac{S_1 - \zeta(1)}{\sqrt{\zeta(l_n)}} < x \right\} P\{\nu_n = 1\}.$$

Let $p>0$ be a number such that qp is an integer. Then

§4.1. Sums of Dependent Indicators

$$P\left\{\frac{S_{\nu_n}-\zeta(\nu_n)}{\sqrt{\zeta(1_n)}}<x\right\} = \sum_{i=0}^{pq} \sum_{\frac{i}{q} \leq \frac{\zeta(1)}{\zeta(1_n)} < \frac{i+1}{q}} P\left\{\frac{S_1-\zeta(1)}{\sqrt{\zeta(1_n)}}<x\right\} P\{\nu_n=1\}$$

$$+ \sum_{i=pq+1}^{\infty} \sum_{\frac{i}{q} \leq \frac{\zeta(1)}{\zeta(1_n)} < \frac{i+1}{q}} P\left\{\frac{S_1-\zeta(1)}{\sqrt{\zeta(1_n)}}<x\right\} P\{\nu_n=1\} = I_1+I_2. \quad (1.18)$$

Consider the first term in (1.18). We shall use the following theorem proved by Ambrosimov (1976).

Theorem B. Let, in a scheme of series, the random variable $\xi = \sum_{i=1}^{n} \eta_i$ and let the function $\zeta(n)$ and probabilities $b_{i_1 \ldots i_r}$ be such that for $r \leq 9\zeta(n)$ as $n \to \infty$

$$\zeta(n) \to \infty, \quad \max_{1 \leq i \leq n} b_i^{2\zeta(n)} \to 0, \quad (1.19)$$

$$\frac{1}{j!} \sum_{I(n)_{m+j}} \left[b_{i_1 \ldots i_{m+j}} + b_{i_1 \ldots i_{m+j}} \right] = o(\zeta^{m-1}(n) e^{-\zeta(n)}) \quad (1.20)$$

uniformly on $m \in \Delta(1,[2\zeta(n)])$ and on $j \in \Delta(0,[9\zeta(n)]-m)$ (here $I_r(n)$ are some exclusive sets) and let

$$\frac{b_{i_1 \ldots i_r}}{b_{i_1} \ldots b_{i_r}} = 1 + o(e^{-2\zeta(n)}), \quad n \to \infty, \quad (1.21)$$

uniformly on $(i_1, \ldots, i_r) \in I_r(n)$ and on $r \in \Delta(0,[9\zeta(n)])$. Then

$$\lim_{n \to \infty} P\left\{\frac{\xi-\zeta(n)}{\sqrt{\zeta(1_n)}} < x\right\} = \Phi(x).$$

It is not difficult to see that

$$\sum_{i=0}^{pq} \sum_{\frac{i}{q} \leq \frac{\zeta(1)}{\zeta(1_n)} < \frac{i+1}{q}} P\left\{\frac{S_1-\zeta(1)}{\sqrt{\zeta(1_n)}} < x\sqrt{\frac{q}{i+1}}\right\} P\{\nu_n=1\} \leq I_1 \leq$$

$$\leq \sum_{i=0}^{pq} \sum_{\frac{i}{q} \leq \frac{\zeta(1)}{\zeta(1_n)} < \frac{i+1}{q}} P\left\{\frac{S_1 - \zeta(1)}{\sqrt{\zeta(1)}} < x\sqrt{\frac{q}{i}}\right\} P\{\nu_n = 1\}. \quad (1.22)$$

If the quantity $1 \to \infty$ such that for any fixed i and q

$$\zeta(1_n)\frac{i}{q} \leq \zeta(1) < \zeta(1_n)\frac{i+1}{q},$$

then S_1 satisfies all the conditions of Theorem B. Therefore

$$\limsup_{n \to \infty} I_1 \leq \sum_{i=0}^{pq} \Phi(x\sqrt{q/i})[T((i+1)/q) - T(i/q)],$$

$$\liminf_{n \to \infty} I_1 \geq \sum_{i=0}^{pq} \Phi(x\sqrt{q/i+1})[T((i+1)/q) - T(i/q)].$$

Since for any fixed p and $q \to \infty$ the last sums tend to the same integral,

$$\lim_{n \to \infty} I_1 = \int_0^p \Phi(x/\sqrt{u}) dT(u). \quad (1.23)$$

The second term in (1.18) has the estimate

$$\sum_{i=pq}^{\infty} P\left\{\frac{i}{q} \leq \frac{\zeta(\nu_n)}{\zeta(1_n)} < \frac{i+1}{q}\right\} \leq P\left\{\frac{\zeta(\nu_n)}{\zeta(1_n)} \geq p\right\},$$

which will be arbitrarily small for any sufficiently large p. Therefore

$$\lim_{n \to \infty} P\left\{\frac{S_{\nu_n} - \zeta(\nu_n)}{\sqrt{\zeta(1_n)}} < x\right\} = \int_0^{\infty} \Phi(x/\sqrt{u}) dT(u).$$

The theorem is thus proved.

As it has been noted, using the results of this section, we will investigate the asymptotical properties of different families of particles in the branching process. However these results can also be used in other problems. For example, using these results, it is possible to obtain limit theorems for the number of cells and the rate of convergence in these theorems for a scheme of the allocation of particles into cells (Kolchin et. al., 1978).

§4.2. FAMILY OF PARTICLES IN CRITICAL PROCESSES

§2a. The Model

Let $\mu(t)$, $t \in N_0$, be a discrete time and homogeneous branching process, let as before $Q_{ij}(t) = P\{\mu(t)=j | \mu(0)=i\}$, and

$$A = \sum_{j=1}^{\infty} j Q_{1j}(1), \quad B = \sum_{j=2}^{\infty} j(j-1) Q_{1j}(1).$$

We will consider the processes $Z_2(\tau,t,d)$ and $\nu_a^b(\tau,t)$, where the first process is defined by relation (0.2) and the second by the relation

$$\nu_a^b(\tau,t) = \sum_{\alpha \in \mathcal{A}_\tau} \chi\{a \leq \mu^{(\alpha)}(t-\tau) < b\}, \quad a, b \in N.$$

It easy to see that $\zeta_k(\tau,t) = \nu_k^{k+1}(\tau,t)$ is the number of particles in $\mathcal{A}_\tau$, having exactly k offspring at time t, and that $\nu_1^\infty(\tau,t) = |\mathcal{B}(\tau,t)|$ is the so-called reduced branching process (the number of particles at time τ having a nonempty set of offspring at time t) considered, for example, by Fleishmann and Siegmund-Schultze (1977) and by Yakimov (1980). Finally we denote by $\eta_m(\tau,t)$ the number of ks, for which $\zeta_k(\tau,t)=m$. The following relation holds for processes $\eta_m(\tau,t)$ and $Z_r(\tau,t,d)$:

$$Z_r(\tau,t,0) = \sum_{m=r}^{\infty} \binom{m}{r} \eta_m(\tau,t). \qquad (2.1)$$

The process $\eta_m(\tau,t)$ has the following interpretation. We decompose the population $\mathcal{A}_\tau$ into groups of particles having the same positive number of offspring at time t. Then $\eta_m(\tau,t)$ is the number of groups containing exactly m particles.

Processes $\zeta_k(\tau,t)$ and $\eta_m(\tau,t)$ can be connected with the problems of allocation of particles into cells. Let τ and t, $\tau < t$, be non-negative integers and let $\mu(t)$, $\mu_i(t)$, $i \in N$, be independent and identically distributed branching processes. Let us assume that at time $\tau+1$ the random number $\mu(\tau)$ particles independently allocate into cells enumerated by $0,1,2,..$ such that the ith particle falls into the kth cell, if $\mu_i(1)=k$. After allocation, the particles can walk along the cells according to the branching processes $\mu_i(t)$, $i \in N$. This means that the ith particle, allocated at time $\tau+1$, will be transferred into a cell with the number $\mu_i(t-\tau)$ at time t. In this interpretation the variable $\zeta_k(\tau,t)$ is the number of particles, allocated at time $\tau+1$ and falling into a cell with number k at time t, and

$\eta(\tau,t)$ is the number of cells containing exactly m particles.
We also put $\mu(0)=1$,

$$Q(t)=P\{\mu(t)>0\}, \quad \gamma(t,d) = \sum_{j=1}^{\infty} Q_{1j}(t) \sum_{\substack{|i-j|\leq d \\ i\geq 1}} Q_{1i}(t),$$

$$B(t)=M\mu(t)(\mu(t)-1), \quad A_2(\tau,t,d) = M[Z_2(\tau,t,d)|\mu(\tau)>0].$$

Then

$$A_2(\tau,t,d) = \frac{B(\tau)}{2Q(\tau)} \gamma(t-\tau,d). \tag{2.2}$$

The relation (2.2) shows that the asymptotical behavior of the process $Z_2(\tau,t,d)$ depends on the local properties of the process $\mu(t)$. Therefore we shall consider the case, where $\mu(t)$ is a non-periodical, (i.e. the greatest common divisor of $\{k:Q_k(1)>0\}=1$), Markovian and discrete-time branching process. The local probabilities of such a process have been studied by Kesten et.al.(1966) in detail. However the arguments used here are applicable to the investigation of more general models of branching processes. To do this it is sufficient to obtain the asymptotics of $\gamma(t,\tau,d)$ as $t,\tau\to\infty$.

§2b. Limit Theorems

Let $l(t)$ be a positive function such that $l(t)\to\infty$, $l^3(t)\ln l(t)=o(\ln t)$ as $t\to\infty$ and let $B(\varphi,\Pi)$ be the class of branching processes from Definition 3.0.1.

$$\theta(m,\tau,t) = M[Z_2(\tau,t,d)|\mu(\tau)=m] = \binom{m}{2}\gamma(t-\tau,d),$$

$$\pi_8(\Delta,x)=\sqrt{\Delta} \sum_{0\leq k\leq x} \frac{1}{2k!} \int_0^\infty u^{k-1/2} e^{-u-\sqrt{u\Delta}} du,$$

$$\pi_9(\Delta,x) = \frac{2}{B} \int_0^\infty \Phi(x/u) e^{-2u/B} du, \quad \lambda(\tau,t) = \frac{t-\tau}{\sqrt[3]{t^2}}.$$

Theorem 2.1. Let $A=1$, $M\mu^2(1)\ln\mu(1)<\infty$ and $t,\tau\to\infty$.

1) If $t-\tau\to\infty$, then $A_2(\tau,t,d) \sim (2d+1)\tau^2/(t-\tau)^3 B$;

2) if $\lambda(\tau,t)\to\infty$, then $\{Z_2(\tau,t,d)|\mu(\tau)>0\} \xrightarrow{P} 0$;

3) if $\lambda(\tau,t)\to C$, where $C\in(0,\infty)$, then

$$\{Z_2(\tau,t,d)|\mu(\tau)>0\} \in B(y, \pi_8(\Delta,x)), \Delta=2BC^3/(2d+1);$$

§4.2. Family of Particles in Critical Processes

4) if $\lambda(\tau,t) \sim 1/l(t)$, then

$$\{Z_2(\tau,t,d)|\mu(\tau)>0\} \in B\left(\frac{y-\theta(\mu(\tau),\tau,t)}{\sqrt{\theta(\tau,t,t)}}, \pi_9(B,x)\right).$$

Remark 2.1. The condition $M\mu^2(1)\ln\mu(1)<\infty$ is connected with the local limit theorems proved by Kesten, Ney, Spitzer (1966) which we shall use for the investigation of asymptotics of $\gamma(\tau,t,d)$.

It follows from the limit theorem for critical processes (§2.1) and Lemma 2.1, which will be proved below, that the following statement is true for the "centering" process in part 4 of Theorem 2.1.

Proposition 2.1. Under the conditions of part 4 of Theorem 2.1

$$P\left\{\frac{\theta(\mu(\tau),\tau,t)}{\sqrt{l^3(t)\theta(\tau,\tau,t)}} < x \Big| \mu(\tau)>0\right\} \longrightarrow E\left(\sqrt{x}\sqrt[4]{8/B(2d+1)}\right),$$

where $x>0$ and $E(x)=1-e^{-x}$.

Lemma 2.1. If $A=1$, $M\mu^2(1)\ln\mu(1)<\infty$ and $B>0$, then

$$\gamma(t,d) \sim \frac{2d+1}{2}\left(\frac{2}{bt}\right)^3, \quad t\to\infty.$$

Proof. Under the conditions of the lemma, it follows from the local limit theorem for critical branching processes (Kesten et.al. (1966)) that, if $j\to\infty$ and $t\to\infty$ such that $j/t\leq\text{const}<\infty$, then

$$Q_{1j}(t) \sim (2/Bt)^2 e^{-2j/Bt}, \quad t\to\infty. \tag{2.3}$$

It is also known that

$$\sup_{j,t} t^2 Q_{1j}(t) < \infty. \tag{2.4}$$

Let $\lambda(t)$ be an integer-valued positive function such that $\lambda(t)\to\infty$, $\lambda(t)=o(t)$, $t\to\infty$, and $\varepsilon>0$. Consider the relation:

$$\gamma(t,d) = \sum_{j=1}^{\lambda(t)-1} Q_{1j}(t) \sum_{\substack{|i-j|\leq d \\ i\geq 1}} Q_{1i}(t) + \sum_{j=\lambda(t)}^{[\varepsilon^{-1}t]} \sum_{\substack{|i-j|\leq d \\ i\geq 1}} Q_{1i}(t) +$$

$$+ \sum_{j=[\varepsilon^{-1}t]+1}^{\infty} Q_{1j}(t) \sum_{\substack{|i-j|\leq d \\ i\geq 1}} Q_{1i}(t) = I_1 + I_2 + I_3. \tag{2.5}$$

IV. Families of Particles in Branching Processes

If we use (2.4), we have the estimate:

$$I_1 = \frac{1}{t^4} \sum_{j=1}^{\lambda(t)-1} t^2 Q_{1j}(t) \sum_{\substack{|i-j|\le d \\ i\ge 1}} t^2 Q_{1i}(t) < \frac{2d+1}{t^4}\sup_{j,t}(t^2 Q_{1j}(t))^2 \lambda(t). \tag{2.6}$$

In order to estimate the second term, we use (2.3) and obtain for any $\varepsilon>0$ as $n\to\infty$ that

$$I_2 \sim (2/bt)^4 \sum_{j=\lambda(t)}^{[\varepsilon^{-1}t]} e^{-2j/Bt} \sum_{\substack{|i-j|\le d \\ i\le 1}} e^{-2i/Bt}.$$

It follows from the inequality

$$(2d+1)e^{-2d/Bt} \le e^{2j/Bt} \sum_{\substack{|i-j|\le d \\ i\le 1}} e^{-2i/Bt} \le (2d+1)e^{2d/Bt}$$

that as $t\to\infty$ for any fixed d

$$I_2 \sim (2d+1)(2/Bt)^4 \sum_{j=\lambda(t)}^{[\varepsilon^{-1}t]} e^{-4j/Bt} \sim \frac{2d+1}{2}(2/Bt)^3(1-e^{-4/B\varepsilon}). \tag{2.7}$$

By virtue of (2.4) and Chebyshev's inequality

$$I_3 \le \frac{2d+1}{t^2}\sup_{j,t}(t^2 Q_{1j}(t))P\{\mu(t)>t/\varepsilon\} \le \varepsilon \frac{2d+1}{t^2}\sup_{j,t}(t^2 Q_{1j}(t)). \tag{2.8}$$

Turning ε to zero in relations (2.7), (2.8) and granting relation $\lambda(t)=o(t)$ in (2.6), we obtain the statement of the lemma from the relation (2.5).

Proof of Theorem 2.1. It is known (see Sevas'yanov (1971)) that

$$Q(\tau) \sim 2/B\tau, \quad B(\tau) \sim B\tau, \quad \tau\to\infty. \tag{2.9}$$

Part 1 of the theorem follows from relation (2.2), Lemma 2.1 and (2.9).

In order to obtain the second part of the theorem, it is sufficient to use Chebyshev's inequality and the asymptotics of $A_2(\tau,t,d)$ from part 1.

Proof of part 3. It is enough to show that the conditions of Theorem 1.1 are fulfilled for the sum

$$Z_2(\tau,t,d) = \frac{1}{2}\sum_{\alpha_{i_1},\alpha_{i_2}\in \mathcal{B}(\tau,t)} \chi(\alpha_{i_1}\ne\alpha_{i_2},\ \rho(\alpha_{i_1},\alpha_{i_2})\le d). \tag{2.10}$$

Since by virtue of the total probability formula

§4.2. Family of Particles in Critical Processes

$$M[\chi(\rho(\alpha_1,\alpha_2)\leq d);\ \alpha_1,\alpha_2 \in (\tau,t)] = \gamma(t-\tau,d),$$

using Lemma 2.1, we obtain:

$$\binom{\mu(t)}{2} M[\chi(\rho(\alpha_1,\alpha_2)\leq d);\ \alpha_1,\alpha_2 \in (\tau,t)] = \frac{2d+1}{2BC^3}\left[\frac{2\mu(\tau)}{b\tau}\right]^2 + W(\tau,t,d), \quad (2.11)$$

where for any $\varepsilon > 0$

$$P\{|W(\tau,t,d)| > \varepsilon | \mu(\tau) > 0\} \to 0 \quad (2.12)$$

as $t \to \infty$, $\tau \to \infty$ and $t-\tau \sim Ct^{2/3}$, by virtue of the limit theorem for critical branching processes §2.1.

On the other hand, using the limit theorem mentioned above once more, we obtain that

$$\lim_{\tau \to \infty} P\left\{\frac{2d+1}{2BC^3}\left[\frac{2\mu(\tau)}{B\tau}\right]^2 < x\ \Big|\ \mu(\tau) > 0\right\} = E\left(\sqrt{\frac{2BC^3 x}{2d+1}}\right). \quad (2.13)$$

It follows from (2.11)-(2.13), that if the conditions of part 3 of Theorem 2.1 are fulfilled, then the conditional distribution of the variable

$$\binom{\mu(t)}{2} M[\chi(\rho(\alpha_1,\alpha_2)\leq d);\ \alpha_1,\alpha_2 \in (\tau,t)],$$

given the $\mu(\tau) > 0$ tends to

$$E\left(\sqrt{\frac{2BC^3 x}{2d+1}}\right)$$

as $\tau, t \to \infty$ and $t-\tau \sim Ct^{2/3}$.

For a sequence $l_\tau \to \infty$ as $\tau \to \infty$, let

$$(\tfrac{1}{2}\tau)M[\chi(\rho(\alpha_1,\alpha_2)\leq d);\ \alpha_1,\alpha_2 \in (\tau,t)] \to \lambda > 0. \quad (2.14)$$

Using the total probability formula, we have:

$$M[\chi(\rho(\alpha_1,\alpha_2)\leq d)\chi(\rho(\alpha_1,\alpha_3)\leq d)] = \sum_{i=1}^{\infty} Q_{1i}(t-\tau)\left[\sum_{|i-j|\leq d} Q_{1j}(t-\tau)\right]^2.$$

for $\alpha_i \in \mathcal{B}(\tau,t)$, $i=1,2,3$. It follows from (2.4) that the sum on the right-hand side of the last relation is not greater than $C_1 Q(t-\tau)(t-\tau)^{-4}$, $C_1 \in (0,\infty)$. Therefore, granting (2.14) and the asymptotics of non-extinction probability, we have:

$$l_\tau^2 M[\chi(\rho(\alpha_1,\alpha_2)\leq d)\chi(\rho(\alpha_1,\alpha_2)\leq d);\ \alpha_i \in \mathcal{B}(\tau,t)] \to 0 \quad (2.15)$$

as $\tau, t \to \infty$ and $t-\tau \sim Ct^{2/3}$. Thus the conditions of Theorem 1.1 are fulfilled. Part 3 of the theorem follows then from Theorem 1.1.

Proof of part 4. We shall prove that Theorem 1.3 is applicable to $Z_2(\tau,t,d)$. Denote by $I=(i_1,i_2)$ and $I_k=(i_{1k},i_{2k})$ the ordered pairs such that $1 \leq i_1 < i_2 \leq \tau$, and denote by $\{I\}$ the set with the same elements of I. We also denote by Δ_τ the set of all pairs I and put

$$\chi_I = \chi(\rho(\alpha_{i_1}, \alpha_{i_2}) \leq d, \ \alpha_{i_1}, \alpha_{i_2} \in \mathcal{B}(\tau,t)),$$

$$b_{I_1 \ldots I_k} = P\{\chi_{I_1} = \ldots = \chi_{I_k} = 1\}, \quad k=1,2,\ldots .$$

Then

$$\theta(\tau,\tau,t) = \sum_{I \in \Delta_\tau} b_I = \binom{\tau}{2} \gamma^2(t-\tau,d), \quad \max_{I \in \Delta_\tau} b_I = \gamma(t-\tau,d).$$

It follows from Lemma 2.1 and the choice of $l(t)$ that, if $t,\tau \to \infty$ and $t-\tau \sim t^{2/3}/l(t)$, then $\theta(\tau,\tau,t) \to \infty$ and $\gamma(t-\tau,d)e^{2\theta(\tau,\tau,t)} \to 0$. Therefore conditions (1.14) of Theorem 1.3 are fulfilled.

We shall accept as exclusive sets $I_r(\tau)$, $r \geq 2$, the set of collections $(I_1, \ldots, I_r)$ for which there exists at least a pair k, $1 \in \Delta(1,r)$, $\{I_k\} \cap \{I_1\} \neq \emptyset$, where $\Delta(m,n) = \{m, m+1, \ldots, n\}$. It is obvious that

$$I_r(\tau) = \bigcup_{l=2}^{2r-1} K_l(\tau),$$

where $K_l(\tau) \subseteq I_r(\tau)$ consists of all collections $(I_1, \ldots, I_r)$ which have exactly l different indexes i_{km}.

Using that $|K_l(\tau)| \leq \binom{\tau}{l}\binom{2l}{r}$, we have:

$$\sum_{(I_1,\ldots,I_r) \in I_r(\tau)} b_{I_1} \ldots b_{I_r} \leq \gamma^r(t-\tau,d) \sum_{l=2}^{2r-1} |K_l(\tau)| \leq$$

$$\leq C_2 \tau^{-1} \theta^r(\tau,\tau,t)(2r)^{2r}. \tag{2.16}$$

If we denote

$$a(j,m,\tau) = \frac{1}{j!} \theta^{1-m}(\tau,\tau,t) e^{\theta(\tau,\tau,t)},$$

then it follows from (2.16) that

§4.2. Family of Particles in Critical Processes

$$a(j,m,\tau) \sum_{I_{j+m}(\tau)} b_{I_1} \ldots b_{I_{j+m}} \le C_3 \tau^{-1} r(\tau,t), \quad (2.17)$$

for $j \in \Delta(0, 9[\theta(\tau,\tau,t)]-m)$, $m \in \Delta(1, 2[\theta(\tau,\tau,t)])$, where

$$r(\tau,t) = \theta(\tau,\tau,t)(18\theta(\tau,\tau,t))^{18\theta(\tau,\tau,t)} e^{2\theta(\tau,\tau,t)}.$$

Consider the sum

$$\sum_{(I_1,\ldots,I_r) \in I_r(\tau)} b_{I_1} \ldots b_{I_r} = \sum_{l=2}^{2r-1} \sum_{(I_1,\ldots,I_r) \in K_1(\tau)} b_{I_1} \ldots b_{I_r}. \quad (2.18)$$

Using the estimation of $b_{I_1 \ldots I_r}$ from the paper of Sevast'yanov (1972), we have:

$$b_{I_1 \ldots I_2} \le \gamma^k(t-\tau, d)[\sup_i Q_{1i}(t-\tau)]^{1-2k} \quad (2.19)$$

for $(I_1, \ldots, I_r) \in K_1(\tau)$ and $1 \le k \le 1/2$. Since $\sup t^2 Q_{ij}(t) < \infty$ and $t - \tau \ge$

$\ge (1-\varepsilon) t^{2/3}/l(t)$ for any $\varepsilon \in (0,1)$ and for sufficiently large τ and t, we obtain from (2.18):

$$\sum_{(I_1,\ldots,I_r) \in I_r(\tau)} b_{I_1} \ldots b_{I_r} \le \sum_{l=2}^{2r-1} \gamma^k(t-\tau,d) \left[\frac{C_4}{(t-\tau)^2}\right]^{1-2k} |K_1(\tau)|$$

$$\le \frac{C_5 l^2(t)}{\sqrt[3]{t}} \theta^r(\tau,\tau,t)(2r)^{2r}.$$

Therefore, for $j \in \Delta(0, 9[\theta(\tau,\tau,t)]-m)$ and $m \in \Delta(1, 2[\theta(\tau,\tau,t)])$,

$$a(j,m,\tau) \sum_{I_{j+m}(\tau)} b_{I_1} \ldots b_{I_{j+m}} \le C_6 \frac{l^2(t)}{\sqrt[3]{t}} r(\tau,t). \quad (2.20)$$

By virtue of the choice of the function $l(t)$

$$\theta(\tau,\tau,t) \ln \theta(\tau,\tau,t) = o(\ln \tau) \quad (2.21)$$

as $t, \tau \to \infty$ and $t-\tau \sim t^{2/3}/l(t)$. Taking into account (2.21) in estimations (2.17) and (2.20), we can see that condition (1.15) of Theorem 1.3 is satisfied.

It is obvious that for our exclusive sets $b_{I_1 \ldots I_r} = b_{I_1} b_{I_2} \ldots b_{I_r}$, when

$(I_1,\ldots,I_r) \notin I_r(\tau)$, $r \geq 2$. This means that condition (1.16) of Theorem 1.3 is also satisfied.

Now let l_τ be a sequence of non-negative integers such that $l_\tau \to \infty$, $\tau \to \infty$, and

$$0 < a \leq \theta(l_\tau, \tau, t)/\theta(\tau, \tau, t) \leq b < \infty.$$

It follows from the definition of $\theta(m,\tau,t)$ that $l_\tau \asymp \tau$ as $\tau \to \infty$. Granting this, we can see that when $\tau, t \to \infty$ and $t-\tau = t^{2/3}/l(t)$, conditions (1.14)–(1.16) of Theorem 1.3 are fulfilled for $\nu_\tau \equiv l_\tau$.

Finally we consider the relation

$$\frac{\theta(\mu(\tau),\tau,t)}{\theta(\tau,\tau,t)} = \left(\frac{\mu(\tau)}{\tau}\right)^2 \frac{\tau}{\tau-1} + \frac{\mu(\tau)}{\tau(\tau-1)}. \quad (2.22)$$

It follows from Chebyshev's inequality that for any $\varepsilon > 0$

$$P\left\{\frac{\mu(\tau)}{\tau(\tau-1)} > \varepsilon \mid \mu(\tau) > 0\right\} \leq (Q(\tau)\tau(\tau-1)^{-1}\varepsilon) .$$

Hence we have from the asymptotic behavior of non-extinction probability that the second term in (2.22) converges to zero as $\tau \to \infty$ in probability.

It follows from the limit theorem for critical processes that the conditional distribution of the first term in (2.22) tends to $E\left(\frac{2}{B}\sqrt{x}\right)$.

Thus the statement of part 4 follows from Theorem 1.3. Theorem 2.1 is thus proved.

The statement of Theorem 2.1 is relative to the case, when $t-\tau \to \infty$. Now we mention a result related to the case of $t-\tau = \text{const}$.

Corollary 2.1. If $A=1$ and $B \in (0,\infty)$, then for $\tau = \text{const} \in N_0$

$$\lim_{t \to \infty} P\{t^{-1}Z_2(t,t+\tau,d) < x \mid \mu(\tau) > 0\} = E\left(\frac{2x}{B\gamma(\tau,d)}\right).$$

The statement of this corollary can be deduced by Lemma 1.1 from the limit theorem for critical processes.

Remark 2.2. Theorem 2.1 and Corollary 2.1 show that the number of pairs of particles having the same number of offspring at the time t is asymptotically equal to zero when τ belongs to the $[0, t-\lambda^*(t)\sqrt[3]{t^2}]$ with $\lambda^*(t) \to \infty$. If τ is nearer to t, this number has several limit distributions depending on the behavior of $t-\tau$.

§4.2. Family of Particles in Critical Processes

Theorem 2.2. Under the conditions of Theorem 2.1

$$P\{Z_2(\tau,t,0)=\eta_2(\tau,t) \mid \mu(\tau)>0\} \to 1.$$

In order to prove this theorem, we need the following lemma.

Lemma 2.2. If $A=1$, $M\mu^2(1)\ln\mu(1)<\infty$, $B>0$, then for $r\in N$

$$\sum_{j=1}^{\infty} Q_{1j}^r(t) \sim \frac{1}{r}\left(\frac{2}{Bt}\right)^{2r-1}, \quad t\to\infty.$$

The proof of this lemma is similar to the proof of Lemma 2.1 and, therefore, it will be omitted.

Proof of Theorem 2.2. We denote

$$H_r = H_r(\tau,t) = \{Z_r(\tau,t,0)=0\}, \quad W(\tau,t) = \eta_2(\tau,t) - Z_2(\tau,t,0).$$

It follows from (2.1) that for any τ, t and $\varepsilon>0$

$$P\{|W(\tau,t)|>\varepsilon|H_3\}=0.$$

Therefore

$$P\{|W(\tau,t)|\varepsilon>|\mu(\tau)>0\} \leq P\{\bar{H}_3|\mu(\tau)>0\}. \tag{2.23}$$

We shall prove that the probability on the right hand side of (2.23) tends to zero. To do this, we represent it in the form

$$P\{\bar{H}_3|\mu(\tau)>0\} = (Q(\tau))^{-1} \sum_{l=3}^{\infty} P\{\bar{H}_3|\mu(\tau)=l\}Q_{11}(\tau). \tag{2.24}$$

Let $\delta\in(0,1)$. Then in the sum

$$\sum_{l=3}^{\infty} P\{\bar{H}_3|\mu(\tau)=l\}Q_{11}(\tau) = \sum_{l=3}^{[B\tau/2\delta]} P\{\bar{H}_3|\mu(\tau)=l\}Q_{11}(\tau)$$

$$+ \sum_{l=[B\tau/2\delta]+1}^{\infty} P\{\bar{H}_3|\mu(\tau)=l\}Q_{11}(\tau) = I_1+I_2 \tag{2.25}$$

the first part has the estimate

$$I_1 = \sum_{l=3}^{[B\tau/2\delta]} P\{\bigcup_{1\leq i_1<i_2<i_3\leq l}\{\chi(\mu_{i_1}(t-\tau)=\mu_{i_2}(t-\tau)=\mu_{i_3}(t-\tau)\neq 0)=1\}\}Q_{11}(\tau)\leq$$

$$\leq 6^{-1}([B\tau/2\delta]+1)^3 Q(\tau) \sum_{i=1}^{\infty} Q_{1i}^3(t-\tau). \tag{2.26}$$

By Chebyshev's inequality we obtain that

$$I_2 \leq 2\delta/B\tau. \tag{2.27}$$

Thus we have from (2.24)-(2.27) that for any $\delta \in (0,1)$ and fixed τ and t

$$P\{\bar{H}_3 | \mu(\tau) > 0\} \le ([B\tau/2\delta]+1)^3 \sum_{i=1}^{\infty} Q_{1i}^3(t-\tau) + 2\delta/B\tau Q(t). \quad (2.28)$$

It follows from Lemma 2.2 that the sum in the first term of (2.28) has the order $(t-\tau)^5$ as $t-\tau \to \infty$. Therefore the first term is equivalent to $C_7 l^5(t)/\sqrt[3]{t}$ when $t-\tau \sim t^{2/3}/l(t)$. But this shows that the first term tends to zero when $l(t)$ satisfies the conditions $l(t) \to \infty$, $\ln l(t) = o(\ln t)$ or $l(t) = o(1)$ as $t \to \infty$. The theorem is proved.

Corollary 2.2. Let $A=1$, $M\mu^2(1)\ln\mu(1) < \infty$ and $t, \tau \to \infty$.

1) If $\lambda(\tau,t) \to \infty$, then $\{\eta_2(\tau,t) | \mu(\tau) > 0\} \xrightarrow{P} 0$;

2) if $\lambda(\tau,t) \to C$, $C \in (0,\infty)$, then

$$\{\eta_2(\tau,t) | \mu(\tau) > 0\} \in B(y, \pi_8(2BC^3, x));$$

3) if $\lambda(\tau,t) \sim 1/l(t)$, then

$$\{\eta_2(\tau,t) | \mu(\tau) > 0\} \in B\left(\frac{y - \theta_o(\mu(\tau),\tau,t)}{\sqrt{\theta_o(\tau,\tau,t)}}, \pi_9(B,x)\right),$$

where

$$\theta_o(m,\tau,t) = \binom{m}{2}\gamma(t-\tau,o), \quad \lambda(\tau,t) = \frac{t-\tau}{\sqrt[3]{t^2}}.$$

Now we mention a limit theorem for the process $\nu_a^b(\tau,t)$ in the case where a and b are fixed.

Theorem 2.3. Let $A=1$ and $B \in (0,\infty)$.

1. If $Q_{11}(1) \ne 1$, $t \to \infty$ and $t-\tau \sim \sqrt{\sigma t}$, where $\sigma \in (0,\infty)$, and $k \in N_o$, then

$$P\{\nu_a^b(\tau,t) = k | \mu(\tau) > 0\} \to \frac{\sigma}{\sum_{i=a}^{b-1} \nu_i + \sigma} \left[\frac{\sum_{i=a}^{b-1} \nu_i}{\sum_{i=a}^{b-1} \nu_i + \sigma}\right]^k,$$

where $\{\nu_i, i \in N\}$ is the stationary measure of the process $\mu(t)$.

2. if $\tau = [t(1-x)]$, $0 < x < 1$, $k \in N_o$, then

$$\lim_{t \to \infty} P\{\nu_1^\infty(\tau,t) = k, | \mu(\tau) > 0\} = x(1-x)^k.$$

Remark 2.3. Since $\zeta_k(\tau,t) = \nu_k^{k+1}(\tau,t)$ it follows from part 1 of Theorem 2.3 that if $t \to \infty$ and $t-\tau \sim \sqrt{\sigma t}$, then

§4.3. Supercritical and Subcritical Processes

$$P\{\zeta_k(\tau,t)=i \mid \mu(\tau)>0\} \to \frac{\sigma}{\nu_k+\sigma}\left(\frac{\sigma}{\nu_k+\sigma}\right).$$

Remark 2.4. From Fleishmann and Siegmund-Schultze's theorem (1977) the conditional limit theorem for $\nu_1^\infty(\tau,t)$ under the condition $\mu(t)>0$ can be deduced.

The proof of Theorem 2.3 is similar to the proof of part 3 of Theorem 2.1 besause it is also based on the use of Theorem 1.1.

§4.3. FAMILIES OF PARTICLES IN SUPERCRITICAL AND SUBCRITICAL PROCESSES

In this section we will prove some limit theorems for the above introduced processes $Z_r(\tau,t,d)$ and $\eta_r(\tau,t)$, $r\geq 2$, in the case, where the initial process $\mu(t)$ is supercritical or subcritical.

It is known that (see Sevas'yanov, 1971, for example) if $A>1$ and $M\mu(1)\ln\mu(1)<\infty$, then $\mu(t)A^{-t}$ converges to a random variable W with probability 1, where the distribution function $\mathcal{K}(x)=P\{W<x\}$ is absolutely continuous for $x>0$ and has a jump q at the point $x=0$, where q is the extinction probability. Thus the distribution $\mathcal{K}(x)$, $x>0$, has a density function $\omega(x)\geq 0$. Assume that in the supercritical case

$$\limsup_{t\to\infty} A^t \, Q_{1j}(t) < \infty. \tag{3.1}$$

If $A<1$, then (see Sevast'yanov (1971))

$$\lim_{t\to\infty} \frac{Q_{1j}(t)}{Q(t)} = P_j^*, \quad \sum_{j=1}^\infty P_j^* = 1,$$

where $Q(t)=P\{\mu(t)>0\}$, and if, in addition, the condition $M\mu(1)\ln\mu(1)<\infty$ is satisfied, then as $t\to\infty$

$$Q(t) \sim KA^t, \quad K\in(0,\infty). \tag{3.2}$$

§3a. Supercritical Processes

We introduce the following notations:

$$D = \int_0^\infty \omega^2(x)dx,$$

$$\pi_{10}(\gamma,x) = \sum_{k\leq x} (k!)^{-1} \int_0^\infty u^k e^{-u} d\mathcal{K}(\sqrt{2A^\gamma u/(2d+1)D}),$$

$$\pi_{11}(x) = \int_0^\infty \phi(x/u) d\mathcal{K}(u), \quad \psi(\tau)=[A^\tau],$$

and denote by $L_r(a,b)$ the class of functions $f(x)$ for which $\int_a^b |f(x)|^r dx < \infty$. We also use the notations from §4.2.

Theorem 3.1. Let $A>1$, $M\mu(1)\ln\mu(1)<\infty$, $D<\infty$, $t, \tau \to \infty$ and condition (3.1) be satisfied.

1) If $t-\tau \to \infty$, then
$$A_2(\tau,t,d) \sim (2d+1)BDA^{3\tau-t}/2A(A-1)(1-q);$$

2) if $t-3\tau \to \infty$, then $Z_2(\tau,t,d) \xrightarrow{P} 0$;

3) if $t=3\tau+\gamma$, $|\gamma|<\infty$ and $\omega(x) \in L_3(0,\infty)$, then $Z_2(\tau,t,d) \in B(y,\pi_{10}(\gamma,x))$;

4) if $t-\tau \to \infty$, $L(t) \to \infty$ and $L(t)A^{L(t)} = o(t)$ with $L(t)=3\tau-t$, then

$$Z_2(\tau,t,a) \in B\left(\frac{y-\theta(\mu(\tau),\tau,t)}{\sqrt{\theta(\psi(\tau),\tau,t)}}, \pi_{11}(x)\right).$$

Remark 3.1. It follows from the results of Athrea and Ney (1970) that condition (3.1) is satisfied, for example, when $\omega(x)$ is bounded and

$$\sum_{k=1}^\infty kq^{k-1}Q_{1k}(1) < A^{-1}.$$

Remark 3.2. From the limit theorem for supercritical processes we obtain that under the conditions of Theorem 3.1

$$\theta(\mu(\tau),\tau,t) \in B\left(\frac{y}{\sqrt{A^{L(t)}\theta(\psi(\tau),\tau,t)}}, \left(\sqrt{x}\sqrt{\frac{2}{D(2d+1)}}\right)\right).$$

Lemma 3.1. If $A>1$, $M\mu(1)\ln\mu(1)<\infty$, $D<\infty$ and condition (3.1) is satisfied, then $\gamma(t,d) \sim (2d+1)DA^{-t}$.

Proof. We use the local limit theorem for supercritical processes from Athrea and Ney (1972). According to this theorem, if $A>1$, $M\mu(1)\ln\mu(1)<\infty$ and j, $t \to \infty$ such that $0<x_1 \leq jA^{-t} \leq x_2 <\infty$, then

§4.3. Supercritical and Subcritical Processes

$$Q_{1j}(t) \sim A^{-t}\omega(jA^{-t}), \qquad (3.3)$$

where $\omega(x)$ is the density function of $K(x)$.

Let $\varepsilon>0$ and

$$R_t^1=\{j:1\leq j<\varepsilon A^t\}, \quad R_t^2=\left\{j:\varepsilon\leq \frac{j}{A^t}\leq \varepsilon^{-1}\right\}, \quad R_t^3=\{j:A^t\varepsilon^{-1}<j\}.$$

Consider the relation:

$$\gamma(t,d) = \sum_{j\in R_t^1} Q_{1j}(t) \sum_{|i-j|\leq d} Q_{1i}(t) + \sum_{j\in R_t^2} Q_{1j}(t) \sum_{|i-j|\leq d} Q_{1i}(t)$$

$$+ \sum_{j\in R_t^3} Q_{1j}(t) \sum_{|i-j|\leq d} Q_{1i}(t) = I_1+I_2+I_3. \qquad (3.4)$$

Using condition (3.1), we have:

$$\limsup_{t\to\infty} I_1 A^t \leq C_1(K(\varepsilon)-q), \quad 0<C_1<\infty. \qquad (3.5)$$

On the other hand, using (3.3), we obtain that

$$\limsup_{t\to\infty} I_2 A^t = \int_\varepsilon^{1/\varepsilon} \omega^2(x)dx. \qquad (3.6)$$

Granting condition (3.3) again, by Chebyshev's inequality we have for any positive ε:

$$I_3 \leq \frac{2d+1}{A^t} \sup_{j,t}(A^t Q_{1j}(t))\, P\{\mu(t)>A^t/\varepsilon\} \leq \varepsilon\, \frac{2d+1}{A^t} \sup_{j,t}(A^t Q_{1j}(t)).$$

Therefore

$$\limsup_{t\to\infty} I_3 A^t \leq C_2\varepsilon. \qquad (3.7)$$

By putting $\varepsilon\to\infty$ in the relations (3.5)-(3.7), and granting that $K(\varepsilon)\to q$, $\varepsilon\to 0$, and condition $D<\infty$, we obtain the statement of the lemma from relation (3.4).

Proof of Theorem 3.1. In order to receive the asymptotics of $A_2(\tau,t,d)$ when $t-\tau\to\infty$, it is sufficient to grant Lemma 3.1, asymptotics $B(\tau) \sim BA^{2\tau}/A(A-1)$ (see Sevast'yanov, 1971, p.33) and the fact that $Q(\tau)\to 1-q$, $\tau\to\infty$, in formula (2.2). The second part of the theorem follows from the first part by using Chebyshev's inequality.

Let us prove part 3. To do this we show that the conditions of Theorem 1.1 are satisfied for the sum

$$Z_2(\tau,t,d) = \frac{1}{2} \sum_{\alpha_{i_1},\alpha_{i_2} \in (\tau,t)} \chi(\alpha_{i_1} \neq \alpha_{i_2}, \rho(\alpha_{i_1},\alpha_{i_2}) \leq d), \qquad (3.8)$$

where the quantities contained in this formula are defined in the beginning of Chapter IV. Since

$$M[\chi(\rho(\alpha_1,\alpha_2)\leq d); \alpha_1,\alpha_2 \in (\tau,t)] = \gamma(t-\tau,d), \qquad (3.9)$$

it is necessary to obtain the limit distribution of the variable

$$\binom{\mu(t)}{2}\gamma(t-\tau,d) = (\mu(\tau)/A^\tau)^2 \frac{(2d+1)D}{2} A^{3\tau-t} + W(\tau,t), \qquad (3.10)$$

where

$$2W(\tau,t) = \mu^2(\tau)A^{\tau-t}[A^{t-\tau}\gamma(t-\tau,d)-(2d+1)D]-\mu(\tau)\gamma(t-\tau,d).$$

If $t=3\tau+\gamma$, then by virtue of the limit theorem for supercritical processes

$$\lim_{\tau\to\infty} P\left\{\left(\frac{\mu(\tau)}{A^\tau}\right)^2 \frac{D(2d+1)}{2} A^{3\tau-t} \leq x\right\} = K\left(\sqrt{\frac{2xA^\gamma}{D(2d+1)}}\right). \qquad (3.11)$$

It follows from (3.11) and Lemma 3.1 that if $t,\tau\to\infty$ and $t=3\tau+\gamma$, then $W(\tau,t)$ converges to zero in probability. Thus, the relation

$$\lim_{\substack{t,\tau\to\infty \\ t=3\tau+\gamma}} P\{\gamma(t-\tau,d)\binom{\mu(t)}{2}\leq x\} = \left(\sqrt{\frac{2xA^\gamma}{D(2d+1)}}\right) \qquad (3.12)$$

shows that condition (1.2) of Theorem 1.1 is satisfied. Let

$$\gamma(t-\tau,d)\binom{l_\tau}{2} \to \lambda > 0 \qquad (3.13)$$

for sequence $l_\tau \to \infty$, $\tau \to \infty$. We can see that the relation

$$M[\chi(\rho(\alpha_1,\alpha_2)\leq d)\chi(\rho(\alpha_1,\alpha_3)\leq d)] = \sum_{i=1}^\infty Q_{1i}(t-\tau)\left[\sum_{|i-j|\leq d} Q_{1j}(t-\tau)\right]^2 \qquad (3.14)$$

is valid for any $(\alpha_1,\alpha_2,\alpha_3)$, $\alpha_i \in B(\tau,t)$. Using condition (3.1), we obtain that the sum on the right-hand side of (3.14) is not greater than $C_3 A^{2(\tau-t)}Q(t-\tau)$. Since it follows from (3.13) that

$$l_\tau^2 A^{\tau-t} \to C_4 \in (0,\infty), \quad l_\tau A^{\tau-t} \to 0,$$

granting the fact that $Q(t-\tau)\to 1-q$, we obtain from (3.13) and (3.14) that

$$l_\tau^3 M[\chi(\rho(\alpha_1,\alpha_2)\leq d)\chi(\rho(\alpha_1,\alpha_3)\leq d), \alpha_i \in (\tau,t)] \to 0$$

as $t,\tau\to\infty$ and $t=3\tau+\gamma$. Hence condition (1.2) is also fulfilled. The statement of part 3 follows from Theorem 1.1.

§4.3. Supercritical and Subcritical Processes

Now we shall prove part 4. Denote by $I=(i_1,i_2)$, $I_k=(i_{1k},i_{2k})$ collections for which $1 \leq i_1 < i_2 \leq \psi(\tau)$, where $\psi(\tau)=[A^\tau]$. The set of such collections we denote by $\Delta_{\psi(\tau)}$ and we put (see proof of Theorem 2.1)

$$\chi_I = \chi(\rho(\alpha_{i_1},\alpha_{i_2}) \leq d, \quad \alpha_{i_1},\alpha_{i_2} \in (\tau,t)),$$

$$b_{I_1 \ldots I_k} = P\{\chi_{I_1} = \ldots = \chi_{I_k} = 1\}, \quad k=1,2,\ldots$$

Since

$$\theta(\psi(\tau),\tau,t) = \binom{\psi(\tau)}{2} \gamma(t-\tau,d), \quad \max_{I \in \Delta_{\psi(\tau)}} b_I = \gamma(t-\tau,d),$$

using Lemma 3.1, we obtain that

$$\theta(\psi(\tau),\tau,t) \to \infty, \quad \max_{I \in \Delta_{\psi(\tau)}} b_I e^{2\theta(\psi(\tau),\tau,t)} \longrightarrow 0 \qquad (3.15)$$

as $t,\tau \to \infty$, $L(t)A^{L(t)}=o(t)$ and $L(t)=3\tau-t$. In fact the first of these relations follows from Lemma 3.1 and condition $L(t) \to \infty$. The second expression is equivalent as $t,\tau \to \infty$ to

$$D(2d+1)A^{L(t)-2\tau} e^{2D(2d+1)A^{L(t)}}(1+o(1)).$$

Granting conditions $L(t) \to \infty$ and $L(t)A^{L(t)}=o(t)$, we obtain that the logarithm of the last expression tends to $-\infty$.

Let $I_r(\psi(\tau))$, $r \geq 2$, be the exclusive sets introduced in the proof of Theorem 2.1, namely, the set of collections $(I_1,\ldots,I_2)$, for which there exists at least a pair $k, l \in \Delta(1,r)$, $\{I_k\} \cap \{I_1\} \neq \emptyset$, where $\Delta(m,n)=\{m,m+1,\ldots,n\}$.

As in the proof of Theorem 2.1, we shall use the following representation of exclusive sets:

$$I_r(\psi(\tau)) = \bigcup_{l=2}^{2^{-1}} K_l(\psi(\tau)),$$

where $K_l(\psi(\tau)) \subseteq I_r(\psi(\tau))$ consists of all collections $(I_1,\ldots,I_r)$ which have exactly l different indexes i_{km}. Taking into account the estimate $|K_l(\psi(\tau))|$ $\leq \binom{\psi(\tau)}{2} \cdot \binom{l}{2}^{\frac{1}{r}}$ we obtain that

$$\sum_{(I_1,\ldots,I_r) \in I_r(\psi(\tau))} b_{I_1} \ldots b_{I_r} \leq \gamma^r(t-\tau,d) \sum_{l=2}^{2r-1} |K_l(\psi(\tau))| \leq C_5 A^{-\tau} \theta^r(\psi(\tau),\tau,t)(2r)^{2r}.$$

Therefore

$$a(j,m,\psi(\tau)) \sum_{I_{j+m}(\psi(\tau))} b_{I_1}\cdots b_{I_{j+m}} \leq C_6 A^{-\tau} R(\tau,t) \quad (3.16)$$

for $j \in \Delta(0, 9[\theta(\psi(\tau),\tau,t)]-m)$, $m \in \Delta(1, 2[\theta(\psi(\tau),\tau,t)])$, where

$$a(j,m,x) = \frac{1}{j!} \theta^{1-m}(x,\tau,t) e^{\theta(x,\tau,t)},$$

$$R(\tau,t) = \theta(\psi(\tau),\tau,t)(18\theta(\psi(\tau),\tau,t))^{18\theta(\psi(\tau),\tau,t)} \exp\{2\theta(\psi(\tau),\tau,t)\}.$$

Consider relation (2.18), replacing τ by $\psi(\tau)$. Using inequality (2.19) and the estimate of $|K_1(\psi(\tau))|$, we obtain from (2.18):

$$\sum_{(I_1,\ldots,I_r) \in I_r(\psi(\tau))} b_{I_1}\cdots b_{I_r} \leq C_6 \theta^r(\psi(\tau),\tau,t)(2r)^{2r} \sum_{l=2}^{2r-1}(l!)^{-1}\left(\frac{C_7 A^{L(t)}}{A^\tau}\right)^{1-2k}$$

Here $k < 1/2$ and, since k and l are integers, $1-2k \geq 1$. Granting that $C_7 A^{L(t)}/A^\tau$ is less than 1, for any sufficiently large τ and t, we have:

$$a(j,m,\psi(\tau)) \sum_{I_{j+m}(\psi(\tau))} b_{I_1}\cdots b_{I_{j+m}} \leq C_8 A^{L(t)-\tau} R(\tau,t). \quad (3.17)$$

Since $\theta(\psi(\tau),\tau,t) \sim C_9 A^{L(t)}$ the right-hand sides of (3.16) and (3.17) tend to zero as $t,\tau \to \infty$ and $A^{L(t)} L(t) = o(t)$. Thus conditions (1.14) and (1.15) of Theorem 1.1 are fulfilled. By virtue of the choice of exclusive sets, condition (1.16) is also satisfied.

Let now $\psi'(\tau) \to \infty$, $\tau \to \infty$, be a sequence of non-negative integers for which

$$0 < a \leq \theta(\psi(\tau),\tau,t) / \theta(\psi'(\tau),\tau,t) \leq b < \infty.$$

It follows from the definition of $\theta(\psi(\tau),\tau,t)$ that $\psi'(\tau) \asymp \psi(\tau)$ as $\tau \to \infty$. Granting this fact, it is easy to see that conditions (1.14)-(1.16) of Theorem 1.3 are fulfilled for $\psi'(\tau)$ when $t,\tau \to \infty$, $L(t) = 3\tau - t$, $L(t) A^{L(t)} = o(t)$.

It follows from the limit theorem for supercritical processes that the distribution of the variable

$$\theta(\mu(\tau),\tau,t)/\theta(\psi(\tau),\tau,t)$$

tends to $\mathcal{K}(\sqrt{x})$ as $t,\tau \to \infty$. But this fact shows that the last condition of Theorem 1.1 is fulfilled. Consequently, part 4 of the theorem follows from Theorem 1.1. Theorem 3.1 is thus proved.

§4.3. Supercritical and Subcritical Processes

Remark 3.3. Theorem 3.1 shows that the number of particles in the population $\mathscr{A}_\tau$, having the same number of offspring at time t is equal to zero asymptotically, if $\tau \in [0, Ct]$. If $\tau \sim 3^{-1}t$, then it has a discrete limit distribution, and if $\tau \sim Ct$ with $3^{-1} < C < 1$, then its limit distribution is absolutely continuous.

Theorem 3.2. Under the conditions of Theorem 3.1

$$P\{Z_2(\tau, t, 0) = \eta_2(\tau, t) | \mu(\tau) > 0\} \to 1.$$

Lemma 3.2. If $A > 1$, $M\mu(1)\ln\mu(1) < \infty$, $\omega(x) \in L_r(0, \infty)$, $r \geq 2$, and condition (3.1) is satisfied, then

$$\sum_{j=1}^{\infty} Q_{1j}^r(t) \sim A^{(1-r)t} \int_{0+}^{\infty} \omega^r(x)\,dx, \quad t \to \infty.$$

The proof of this lemma is similar to the proof of Lemma 3.1.

Proof of Theorem 3.2. Consider relation (2.23). Represent the probability on the right-hand side of (2.23) in the form:

$$P\{\bar{H}_3 | \mu(\tau) > 0\} = \frac{1}{Q(\tau)} \sum_{l=3}^{\infty} P\{\bar{H}_3 | \mu(\tau) = l\} Q_{1l}(\tau). \tag{3.18}$$

Let $\delta \in (0,1)$. Dividing the sum in (3.18) into two parts, from 3 to $[A^\tau \delta^{-1}]$ and from $[A^\tau \delta^{-1}] + 1$ to infinity, we obtain that the first part is not greater than

$$6^{-1}([A^\tau \delta^{-1}] + 1)^3 Q(\tau) \sum_{i=1}^{\infty} Q_{1i}(t-\tau).$$

The second part can be estimated by the probability

$$P\{\mu(\tau) = A^\tau \delta^{-1}\} \leq \delta.$$

Thus for any $\delta \in (0,1)$ and fixed τ and t we have:

$$P\{\bar{H}_3 | \mu(\tau) > 0\} \leq ([A^\tau \delta] + 1)^3 \sum_{i=1}^{\infty} Q_{1i}^3(t-\tau) + \delta. \tag{3.19}$$

It follows from Lemma 3.2 that the first term on the right-hand side of (3.19) has the order $O(A^{L(t)-t})$. Hence, under the condition $L(t)A^{L(t)} = o(t)$, the probability $P\{\bar{H}_3 | \mu(\tau) > 0\}$ tends to zero. Granting this in relation (2.23) we obtain the statement of Theorem 3.2.

§3b. Subcritical Processes

Finally we present a theorem for subcritical processes. Introduce the following notations:

$$W_\tau^{(r)}(m) = \sum_{1 \leq i_1 < \ldots < i_r \leq m} \chi(\mu_{i_1}(\tau) = \ldots = \mu_{i_r}(\tau)\; 0),$$

$$P_k^{(r)}(\tau) = P\{W_\tau^{(r)}(\xi) = k\}, \quad \bar{P}_k^{(r)}(\tau) = P\{W_\tau^{(r)}(\xi) = k \mid \{W_\tau^{(r+1)}(\xi) = k\},$$

where ξ is a random variable with the distribution $\{P_j^*, i \in N\}$.

Theorem 3.3. Let $A < 1$, and $M\mu(1)\ln\mu(1) < \infty$.

1) If $t \to \infty$ and $t - \tau \to \infty$, then $Z_r(\tau, t, 0) \xrightarrow{P} 0$ and $\eta_r(\tau, t) \xrightarrow{P} 0$;

2) if $0 < \tau < \infty$ and $k \in N_0$, then

$$\lim_{t \to \infty} P\{Z_r(t, t+\tau, 0) = k \mid \mu(t) > 0\} = P_k^{(r)}(\tau),$$

$$\lim_{t \to \infty} P\{\eta_r(t, t+\tau) = k \mid \{Z_{r+1}(t, t+\tau) = 0,\; \mu(t) > 0\} = \bar{P}_k^{(r)}(\tau).$$

The proof of this theorem is based on the limit theorem for subcritical processes and is similar to the proofs of the theorems proved above.

REFERENCES

Aldous, D. (1978). Weak convergence of randomly indexed sequences of random variables. Math. Proc. Cambridge Philos. Soc. 83. part 1, 117-126.

Alyev, S.A., Shurenkov, V.M. (1982). Limit phenomena and convergence of Galton-Watson processes to Jirina process. Theory Probab. Appl., 27 (3), 443-455, (Russ.).

Ambrosimov, A.S. (1976). The normal law in the scheme of sums of dependent random variables considered by B.Sevast'yanov. Theory Probab. Appl., 21(1), 184-189, (Russ).

Asmussen, S., Hering, H. (1983). Branching Processes. Boston et. al. Birkhauser. 461p.

Asmussen, S., Hering, H. (1976). Strong limit theorems for supercritical immigration-branching processes. Math. Scand., 39(2), 327-342.

Athreya, K.B., Ney, P. (1972). Branching Processes. Berlin et. al. Springer. XI. 287p.

Athreya, K.B., Ney, P.(1970). The local limit theorem and some related aspects of supercritical branching processes. Trans-Amer. Math. Soc., 152(1), 233--251

Athreya, K.B., Parthasarythy, P.R.,Sankaranarayanan, G.(1974). Supercritical age-dependent branching processes with immigration. J. Appl. Probab. 11(4), 695-702.

Badalbayev, I. S., Rahimov, I. (1978). Critical branching processes with immigration decreasing intensity. Theory Probab. Appl., 23(2), 259-268. (Russ; English trans.).

Badalbayev, I. S. (1982). Limit theorems for critical multitype branching processes with immigration decreasing intensity. In: "Limit Theorems for Random Processes". Tashkent, "Fan"' 41-54,(Russ.).

Badalbayev, I.S., Rahimov, I. (1983). Further results on branching processes with immigration decreasing intensity. Theory Probab. Appl., 28(4), 811--816, (Russ; English trans.).

References

Badalbayev, I.S., Zubkov, A.M.(1983). The limit theorem for the sequence of branching processes with immigration. Theory Probab. Appl. 1983, 28(3), 382-388.

Badalbayev, I.S., Rahimov, I. (1985). New limit theorems for multitype branching processes with decreasing immigration. Izvestiya of AS of Uzbek SSR, 2, 17-22.

Bahr, Von B. (1972). On sampling from a finite set of independent random variables. Z. Wahrchein. Verw. Geb., 24(4), 279-286.

Barbour, A.D., Eagleson, G. K. (1983). Poisson approximation for some statistics on exchangeable trials. Adv. Appl.Probab. 1983, 15(3), 585-600.

Batyrov, Kh., Manevich, D., Nagayev, S. (1977). On Esseen's inequality for random number random variables. Mat. Zametki, 1977, 22(1), 143-146, (Russ).

Bellman, R., Harris, T.E.(1948). On the theory of age-dependent stochastic branching processes. Proc. Nat. Acad. Sci. U.S.A., 34(12), 601-604.

Belyaev, Yu. K. (1975). Probability Methods of Testing Control. Nauka, Moscow, (Russ).

Beska, M., Klopotowski, A., Slominski, L. (1982). Limit theorems for random sums of dependent d-dimensional random vectors Z. Wahrch. Verw. Gebiete, 61, 43-57.

Bikalis, A. (1966). Estimations of the remainder term in the central limit theorem. Litvsk. Math. Sb., 6(3), 323-346.

Borovkov, A. A. (1986). Probability Theory. Nauka, Moscow, (Russ).

Borovkov, K.A. (1988). On a method for proof of limit theorems for branching processes. Theory Probab. Appl. 33(1), 115-123, (Russ).

Brown, B. M. (1971). Martingale central limit theorem. Ann. Math. Statist. 42(1), 59-62.

Brown, B. M., Eagleson, G.K. (1971). Martingale convergence to infinitely divisible laws with finite variances. Trans. Amer. Math. Soc., 162, 449-453.

Chistyakov, V. P. (1957). Local limit theorems of the theory of branching random process. Theory Probab. Appl. II(3), 360-374, (Russ).

Csorgo, M., Rychlik, Z. (1980). Weak convergence of sequences of random elements with random indices. Math. Proc. Cambridge Plilos. Soc. 88. part 1, 171-174.

Dorogov, V.I., Chistyakov, V.P.(1988). Probability models of transformations of particles. Nauka. Moscow, (Russ.).

References

Durham, S.D. (1971). A problem concerning generalized age-dependent branching processes with immigration. Ann. Math. Stat. 42(3), 1121-1123.

Dvoretzky, A. (1972). Asymptotic normality for sums of dependent random variables. Proc. Sixth Berkeley Symp. Math. Statist. 11, 513-536.

Eagleson, G. K. (1975). Martingale convergence to mixtures of infinitely divisible laws. Ann. Probab. 3(3), 557-562.

Erdos, P., Renyi, A. (1959). On the central limit theorem for samples from a finite population. Publ. Math. Inst. of Hungar. Acad. Scienc. 4(1), 49-61.

Esty, W. W. (1975). Critical age-dependent branching processes. Ann. Probab. 3(1), 49-60.

Feller, W. (1967). An Introduction to Probability Theory and its Applcations. II. "Mir". Moscow.

Fleishmann, K., Siegmund - Schultze R. (1977). The structure of reduced critical Galton-Watson processes. Math. Nachr. 79, 233-241.

Foster, J. H. (1971). A limit theorem for a branching process with state -dependent immigration. Ann. Math. Statist. 42(5), 1773-1776.

Foster, J.H., Williamson, J. A. (1971). Limit theorems for the Galton-Watson process with time-dependent immigration. Z. Wahrsch. und Verw. Geb. 20(3), 227-235.

Gaenssler, P., Strobel, J., Stute, W. (1978). On central limit theorems for martingale triangular arrays Acta Math. Acad. Sci. Nungar. 31, 205-216.

Green, P. J. (1977). Conditional limit theorems for general branching processes. J. Appl. Probab. 14(3), 451-463.

Harris, T. E. (1948). Branching processes. Ann. Statist. 19(4), 474-494.

Harris, T. E. (1966). The Theory of Branching Processes. "Mir", Moscow.

Ibragimov, R. (1972). On limit theorems for branching random processes. In "Random Processes and Stat. Probab.", "Fan", Tashkent, 67-72.

Jacod, J., Shiryaev, A.N. (1987). Limit Theorems for Stochastic Processes. Springer-Verlag.

Jagers, P. (1975). Branching Processes with Biological Applications. London. e. a. J. Wiley Sons. XIII.

Jirina, M. (1958). Stochastic branching processes with continuous state space Srechose. Math. J., 8, 292-312.

Jirina, M. (1987). Limit Theorem for sums of independent random variables observed on a finite population. Indian J. of Math., 29(1), 65-83.

Kaverin, S. V. (1991). A refinement of limit theorems for critical branching processes. Theory Probab. Appl. 35(3), 574-580, (Russ. English trans.).

References

Kawazu, K., Watanabe, S. (1971). Branching processes with immigration and related limit theorems. Theory Probab. Appl., 6(1), 34-51.

Kendall, D. G. (1966). Branching processes since 1873. J. London Math.Soc. 41(1), 385-406.

Kesten, H., Ney, P., Spitzer, F. (1966). The Galton-Watson process with mean one and finite variance. Theory Probab. Appl. 11(4), 579-611.

Klein, B., Macdonald, M. (1980). The multitype continuous - time Markov branching processes in a periodic environment. Adv. Appl. Prob., 12(1).

Klopotowski, A. (1977). Limit Theorems for Sums of Dependent Random Vectors in R^d. Dissert. Math. C. L. J. P. 1-62. PWN. Warszawa.

Klopotowski, A. (1980). Mixtures of infinitely divisible distributions as limit laws for sums of dependent random variables. Z. Wahrch. Verw. Geb., 51(1-3), 101-113.

Kolchin, V.F., Sevast'yanov, B.A., Chistyakov, B. P. (1978). Random Allocations. Wiley, New York.

Kolmogorov, A.N. (1938). To the solution of one biological problem. Informations of R. I. of Tomsk University, 7-12.

Kolmogorov, A.N., Dmitriev, N. A. (1947). Branching stochastic processes. Dokl. Akad. Nauk SSSR, 56, 7-10.

Kolmogorov, A.N., Sevast'yanov, B.A. (1947). The calculation of final probabilities of branching random processes. Dokl. Akad. Nauk SSSR, 56, 783-786.

Korolev, V.Yu. (1989). Approximation of distribution of the sum of random number independent random variables by mixtures of normal distributions. Theory Prob. Appl. 34(3), 581-588.

Kruglov, V.M., Korolev, V. Yu.(1990). Limit Theorems for Random Sums. Moscow State University, Moscow, (Russ.).

Kubaski, K. S. (1983). On a random-sums limit problem. Proc. of the 4-th Papponian symp. on Math. Statist. Bad. Tatzm. Austria, 231-263.

Kubaski, K.S., Szynal, D. (1985). Weak convergence of randomly indexed sequences of random variables. Bulletin Polich Acad. Sciences Mathem. 32 (3-4).

Lamperti, J. (1967a). Limiting distributions for branching processes Proc 5-th Berkeley Sympos. Math. Statist. and Probab. 1965-1966., 2, Part 2. Berkeley Los Angeles. 225-241.

Lamperti, J.(1967b). The limit of sequence of branching processes Z. Wahrschein und Verw. Geb. 7, 271-288.

References

Liptser, R.S., Shiryaev, A. N. (1986) Theory of Martingales, Nauka, Moscow, (Russ.).

Mellein, B. (1982). Local limit theorems for the critical Galton - Watson process with immigration. Reista Colombiana de Matematicas. 16, (1-2), 31-56.

Mitov, K.V., Yanev, N.M. (1984). Critical Galton-Watson processes with decreasing state-dependent immigration. J. Appl. Probab. 21(1), 22-39.

Mitov, K.V., Janev, N.M.(1985). Branching processes with decreasing state-dependent immigration. Serdika. 11(1), 25-41.

Mode, C.J. (1971). Multitype Branching Processes: Theory and Applications. New York. Amer. Elsevier. XX.

Nagayev, S.V., Khan, L. (1980). Limit theorems for critical Galton-Watson process with migration. Theory Probab. Appl. 25, 523-534.

Nagaev, S. V. (1975). The limit theorem for branching processes with immigration. Theory Prob. Appl., 20(1), 178-180.

Nagaev, S.V., Asadullin, M.(1985). On certain scheme for summation of random number independent random variables with applications to the branching processes. Dokl. Akad. Nauk SSSR, 285(2), 293-296.

Nerman, O. (1984). The stable pedigrees of critical branching populations J. Appl. Probab. 21(3), 447-463.

Pakes, A.G.(1971a). Some limit theorems for the total progeny of a branching process. Adv. Appl. Probab. 3(1), 176-192.

Pakes, A.G. (1971b). A branching process with a state - dependent immigration component. Adv. Appl. Probab. 3(2), 301-314.

Pakes, A. G. (1972). A limit theorem for the integral of a critical age-dependent branching process. Math. Biosci. 13 (1), 109-112.

Pakes, A. G. (1975). Some results for non-supercritical Galton-Watson processes with immigration. Math. Biosci. 24 (1,2), 71-92.

Pakes, A. G. (1976). Some new limit theorems for the critical branching process allowing immigration. Stoh. Proc. Appl. 4(2), 175-185.

Pakes, A. G. (1986). Some properties of a branching process with group immigration and emigration. Adv. Appl. Probab. 18, 628-645.

Petrov, V.V. (1972). Sums of Independent Random Variables. Nauka. Moscow.

Prokhorov, Ju.V. (1960). On Kolmogorov's uniform limit theorem. Theory Probab. Appl., 5(1), 103-113.

Rahimov, I. (1982). On the branching random processes with immigration in a periodic environment. Abst. of Comm. IV USSR-Japan Symposium on Probab. Theory and Math. Statist. II, Tbilisi, 1982, 2 167-168, (English).

References

Rahimov, I. (1984a). On the limit theorems for a sequence of branching processes with non-homogeneous immigration. Theory Probab. Appl. 29(4), 316-317, (Russ., English trans.).

Rahimov, I. (1984b). Limit theorems for total progeny of particles in critical Galton-Watson processes with immigration. In the book: "Asymptot. Problems of Probability Distributions" "Fan", Tashkent, 106-119 (Russ)

Rahimov, I. (1985a). Convergence of sequence of branching processes with immigration to the Jirina process, In the book "Limit Theorems for Probability Distributions", "Fan", Tashkent, 134-148, (Russ.).

Rahimov, I. (1985b). Limit distributions for integrals of Bellman - Harris processes with non-homogeneous immigration, Informations of AS of Uzbek SSR, No5, 20-28, (Russ.).

Rahimov, I. (1986a). Asymptotics of local probability of the Galton-Watson processes with decreasing immigration, I, II, Informations of AS of Uzbek SSR, No 2, 33-38, No3, 38-43, (Russ.).

Rahimov, I. (1986b). Critical branching processes with infinite variance and decreasing immigration. Theory Probab. Appl. 31(1), 88-100. (Russ., English trans.)

Rahimov, I. (1986c). Sums of dependent indicators and branching stochastic processes. 1st World Congress of Bernoulli Society., 1, 132, Tashkent, (English).

Rahimov, I. (1987a). Limit theorems for random sums of dependent indicators and their applications in the theory of branching processes. Theory Probab. Appl. 32(2), 317-326, (Russian; English trans).

Rahimov, I.(1987b). Limit theorems for decomposable branching processes with immigration. Dokl. of AS of Uzbek SSR, 6, 26-30, (Russian).

Rahimov, I. (1988a). Local limit theorem for branching processes with immigration. "Serdica Bulg. Math.Publ.", 14(2), 234-244, (Russ.).

Rahimov, I.(1988b). Local limit theorems for branching random processes with decreasing immigration. Theory Probab. Appl. 33(2), 387-392, (Russ.,Eng.).

Rahimov, I. (1989a). Asymptotic behavior of families of particles in branching random processes. Soviet Math. Dokl., 39(2), 322-325, (English).

Rahimov, I. (1989b). The methods of summation of random variables in the theory of branching processes. Vth International Vilnius Conference on Theory of Probab. and Math. Statist., Vilnius, 123, (English).

Rahimov, I. (1989c). Asymptotic behavior of families of particles in branching random processes, In the book "Functionals of Random Processes and Statist., "Fan", Tashkent, 146-163, (Russ.).

References

Rahimov, I.(1992a). Sampling sums of dependent random variables, Mixtures of infinitely divisible laws and Branching random processes, Discrete Mathem. and Applications. 2(3), 337-356, (English); Discrete Mathematics (1991) 3(2), 236-257, (Russ.).

Rahimov, I. (1992b). General branching processes with reproduction-dependent immigration Theory Probab. Appl., 37(3), 513 - 525, (Russ., English trans.).

Rahimov, I., Kaverin, S. (1986) The class of limit distributions of critical branching processes with state-dependent immigration, Dokl. of AS of Uzbek SSR, 1, 21-25, (Russ.).

Rahimov, I., Kurbanov, S.(1989). Branching processes with non-homogeneous migration and infinite variance, In the book "Functionals of Random Processes and Statist.", "Fan", Tashkent, 71-85, (Russ).

Rahimov, I., Sirazhitdinov, S. Kh. (1989). Approximation of distribution of a sum in a certain scheme for summation of independent random variables. Soviet Math. Dokl. 38(1), 23-27, (English).

Rahimov, I., Sirazhitdinov, S. Kh. (1988). The limit distribution of the sum in a certain scheme for summation of independent random variables, In the book: "Asymptotical Methods of Probability Theory and Math. Statist.", "Fan", Tashkent, 231-147, (Russian).

Rychlik, Z. (1979). Martingale random central limit theorems. Acta Math. Acad. Sci. Nungar. 34, 129-139.

Sagitov, S. M. (1983). The limit theorem for critical general branching process. Mathem. Zametki, 34(3), 453-461.

Sato, M. (1975). On a Galton-Watson process with state-dependent imigration. Sci. Repts. Niigata Univ., 12, 33-42.

Seneta, E. (1970). On the supercritical Galton-Watson prosesses with immigration. Math. Biosci. 7(1), 9-14 .

Seneta, E. (1985). Regularly Varying Functions, Nauka, Moscow.

Sevast'yanov, B. A. (1951). The theory of branching random process. Uspekhy Math. Nauk, 6(6), 47-99.

Sevast'yanov, B. A. (1972). The Poisson limit law in a scheme of sums of dependent random variables. Theory Probab. Appl., 17(4), 733-738.

Sevast'yanov B. A. (1971). Branching Processes, Nauka, Moscow.

Sevast'yanov, B.A.,Zubkov, A.M. (1974). The regulative branching processes. Theory Probab. Appl., 19(1), 15-25 .

Shiryaev, A. N. (1980). The Probability. "Nauka", Moscow.

Shurenkov, V. M. (1979). On the additive functionals of branching processes. Theory Probab. Appl., 24(2), 389-394.

Shurenkov, V.M. (1976). Two limit theorems for critical branching processes. Theory Probab. Appl., 21(3), 548-558.

Silverman, B., Brown, T. (1978). Short distances, flat trangles and Poisson limits. J. Appl. Probab., 15, 815-825.

Sirazdinov, S. Kh., Mamatov, M.,Formanov, Sh. (1970). Uniform estimations in limit theorems for sums of random number independent random variables. Informations of AS of Uzbek SSR, 5, 28-34.

Slack, R.S. (1968). A branching process with mean one and possibly infinite Variance. Z. Wahrsch. Verb. Geb. 9(2), 139-145.

Sugiman, Ikuo. (1986). A random CLT for dependent random variables J. Maltivariate Analysis. 20, 321-326.

Topchii, V. A. (1982). The local limit theorem for critical Bellman-Harris process. In "Limit Theor. Probability Theory", Novosibirsk, 97-122.

Uchaykin, V.V., Ryzov, V.V. (1988). The Stochastical Theory of the Transfer of High Energy Particles. Nauka, Novosibirsk.

Vatutin, V.A., Zubkov, A. M. (1985) Branching Processes, 1, In: Progress in Science and Technology: Probability Theory . Mathematical Statistics. Theoretocal Cybernetics, 23, 3-67.

Vatutin, V. A. (1976). The critical multitype Bellman-Harris process with immigration. Theory Probab. Appl. 21(2), 447-454.

Vatutin, V. A., Televinova, T. M., Chistyakov, B. P. (1985). Probability Methods in Physical Researchs. Nauka, Moscow, (Russ.).

Vatutin, V.A., Saqitov, S. M. (1988). Decomposable critical branching Bellman-Harris processes with particles of two different types I, II. Theory Probab. Appl. 33(3), 495-507, (1989), 34(2), 251-262, (Russ.,Engl.)

Weiner, H. J. (1972). A critical age-dependent branching process with immigration. Ann. Math. Statist. 43(6), 2099-2103.

Weiner, H.J. (1975). Conditional moments in a critical age-dependent branching process. J. Appl. Probab. 12(3), 581-587.

Yaglom, A. M. (1947). Some limit theorems of the theory of branching processes. Dokl. Akad. Nauk SSSR, 56(8), 795-798.

Yakimov, A. L. (1980).Reduced branching processes. Theory Probab.Appl. 1980, 25(3), 593-596.

Yamazato, M. (1975). Some results on continuous time branching process with state-dependent immigration. J. Mat. Soc. Japan 27(3), 479-496.

Yanev, N.M., Mitov, K.V.(1985). A critical branching process with decreasing migration. Serdica. 11(3), 240-244.

Yanev, N. M., Mitov, K. V. (1982). Limit theorems for controlled branching processes with non-homogeneous migration. Compets Rends de L'Academie Bulgare des Sciences. 35(3), 299-301.

Yermakov, C.M., Nekrutkin, V. V., Sipin, A. A. (1984). Random Processes for the Solution of Classical Equations of Mathematical Physics. Nauka, Moscow, (Russ.).

Zolotorev, V.M. (1957). Revisions of some theorems of the theory of branching processes. Theory Probab. Appl., 2(2), 256-266.

Zubkov, A. (1974). The analogy between the Galton-Watson processes and φ-branching ones. Theory Probab. Appl., 19(2), 319-339.

Zubkov, A. M. (1972). Life periods of branching processes with immigration. Theory Prob. Appl. 17(1), 179-188.

Zubkov, A. M. (1977). Inequalities for the distribution of the sum of functions of independent random variables. Mathem. Zametki, 22(5), 745-758.

Inst. of Mathematics of the Ukrainian A. S. (1978). The Handbook on the Probability Theory and Math. Statist. "Naukova dumka", Kiev.

INDEX

absolutely continuous, 177
accompanying distribution , 110
allocation, 166 ,167
ancestor, 59
Anskombe condition, 19

basic lemma, 50,
basic sample space, 60
Bellman-Harris process, 44,74
birth time, 61

cascade process, 1
centering process, 169
chemical process, 1
childbearing process, 60
collection, 181
compact subset, 144
compensator, 31
continuous theorem, 110,116
continuous time process, 46,54

decreasing immigration, 105
decreasing migration, 122
dependent indicators, 157
dependent variables, 9
direct Riemann integral, 72
distance, 157
double stopping time, 61,77,79

equiprobable scheme, 17
exclusive set, 163,172,173

expectation, 55
exponential distribution, 48
extendent Anscombe condition, 21
extinction probability, 47,177

family of particles, 167
fixed state, 154

Galton-Watson process, 44,105,122
general process, 45,76,96
generating function, 45,48,56
Green function, 106
gamma-distribution, 52,86,125
Gauss function, 125

homogeneous immigration, 54

immigration process, 2
increasing immigration, 79,124
indicator function, 7,122,157
individual characteristic, 71
infinite variance, 124
infinitely divisible, 7

Jirina process, 93

Kramp-Mod-Jagers process, 45,61,95

life-length, 59,63
lifetime, 95
local probability, 110,136

local limit theorem, 136,178
L_1-norm, 138

Markov branching process, 44
mean square convergence, 57
mixture, 6
moment, 46
multiple sum, 9

non-extinction probability, 48,174
non-periodical process, 137,168
non-uniform estimation, 39
non-homogeneous, 105

partial process, 109,110
point process, 59
Poisson process, 2
population process, 2,64
probability space, 60
process with selection, 64
product of σ-algebras, 9
phenomenon, 110

random characteristic, 60,63,74
random environment, 60,61,62
random medium, 58,72
random sum, 6,157
randomly indexed sequence, 18
randomly stopped immigration, 87
reduced process, 167
reproduction process, 59,64,74
regularly varying
       function, 106,109,127
sampling sums, 5
sampling without replacement, 17
scheme of series, 58,64
semimartingale, 30,32

Sevast'yanov process, 44
slowly varying
       function, 95,109
stable distribution, 125
state-dependent immigration, 122
stationary measure, 106
stationary limit
       distribution, 56
stochastically
       continuous, 95,98,100,103
stochastic exponent, 31,32
stochastic exponent method, 5
stopping time, 6,7
switching time, 87

transient probability, 97
transfer theorem, 58
transformation, 1
triplet, 30,
total progeny, 135

variation distance, 158
variance, 55
uniform convergence
       theorem, 128,131
uniform deviation, 38

Lecture Notes in Statistics

For information about Volumes 1 to 8
please contact Springer-Verlag

Vol. 9: B. Jørgensen, Statistical Properties of the Generalized Inverse Gaussian Distribution. vi, 188 pages, 1981.

Vol. 10: A.A. McIntosh, Fitting Linear Models: An Application of Conjugate Gradient Algorithms. vi, 200 pages, 1982.

Vol. 11: D.F. Nicholls and B.G. Quinn, Random Coefficient Autoregressive Models: An Introduction. v, 154 pages, 1982.

Vol. 12: M. Jacobsen, Statistical Analysis of Counting Processes. vii, 226 pages, 1982.

Vol. 13: J. Pfanzagl (with the assistance of W. Wefelmeyer), Contributions to a General Asymptotic Statistical Theory. vii, 315 pages, 1982.

Vol. 14: GLIM 82: Proceedings of the International Conference on Generalised Linear Models. Edited by R. Gilchrist. v, 188 pages, 1982.

Vol. 15: K.R.W. Brewer and M. Hanif, Sampling with Unequal Probabilities. ix, 164 pages, 1983.

Vol. 16: Specifying Statistical Models: From Parametric to Non-Parametric, Using Bayesian or Non-Bayesian Approaches. Edited by J.P. Florens, M. Mouchart, J.P. Raoult, L. Simar, and A.F.M. Smith, xi, 204 pages, 1983.

Vol. 17: I.V. Basawa and D.J. Scott, Asymptotic Optimal Inference for Non-Ergodic Models. ix, 170 pages, 1983.

Vol. 18: W. Britton, Conjugate Duality and the Exponential Fourier Spectrum. v, 226 pages, 1983.

Vol. 19: L. Fernholz, von Mises Calculus For Statistical Functionals. viii, 124 pages, 1983.

Vol. 20: Mathematical Learning Models — Theory and Algorithms: Proceedings of a Conference. Edited by U. Herkenrath, D. Kalin, W. Vogel. xiv, 226 pages, 1983.

Vol. 21: H. Tong, Threshold Models in Non-linear Time Series Analysis. x, 323 pages, 1983.

Vol. 22: S. Johansen, Functional Relations, Random Coefficients and Nonlinear Regression with Application to Kinetic Data, viii, 126 pages, 1984.

Vol. 23: D.G. Saphire, Estimation of Victimization Prevalence Using Data from the National Crime Survey. v, 165 pages, 1984.

Vol. 24: T.S. Rao, M.M. Gabr, An Introduction to Bispectral Analysis and Bilinear Time Series Models. viii, 280 pages, 1984.

Vol. 25: Time Series Analysis of Irregularly Observed Data. Proceedings, 1983. Edited by E. Parzen. vii, 363 pages, 1984.

Vol. 26: Robust and Nonlinear Time Series Analysis. Proceedings, 1983. Edited by J. Franke, W. Härdle and D. Martin. ix, 286 pages, 1984.

Vol. 27: A. Janssen, H. Milbrodt, H. Strasser, Infinitely Divisible Statistical Experiments. vi, 163 pages, 1985.

Vol. 28: S. Amari, Differential-Geometrical Methods in Statistics. v, 290 pages, 1985.

Vol. 29: Statistics in Ornithology. Edited by B.J.T. Morgan and P.M. North. xxv, 418 pages, 1985.

Vol 30: J. Grandell, Stochastic Models of Air Pollutant Concentration. v, 110 pages, 1985.

Vol. 31: J. Pfanzagl, Asymptotic Expansions for General Statistical Models. vii, 505 pages, 1985.

Vol. 32: Generalized Linear Models. Proceedings, 1985. Edited by R. Gilchrist, B. Francis and J. Whittaker. vi, 178 pages, 1985.

Vol. 33: M. Csörgo, S. Csörgo, L. Horváth, An Asymptotic Theory for Empirical Reliability and Concentration Processes. v, 171 pages, 1986.

Vol. 34: D.E. Critchlow, Metric Methods for Analyzing Partially Ranked Data. x, 216 pages, 1985.

Vol. 35: Linear Statistical Inference. Proceedings, 1984. Edited by T. Calinski and W. Klonecki. vi, 318 pages, 1985.

Vol. 36: B. Matérn, Spatial Variation. Second Edition. 151 pages, 1986.

Vol. 37: Advances in Order Restricted Statistical Inference. Proceedings, 1985. Edited by R. Dykstra, T. Robertson and F.T. Wright. viii, 295 pages, 1986.

Vol. 38: Survey Research Designs: Towards a Better Understanding of Their Costs and Benefits. Edited by R.W. Pearson and R.F. Boruch. v, 129 pages, 1986.

Vol. 39: J.D. Malley, Optimal Unbiased Estimation of Variance Components. ix, 146 pages, 1986.

Vol. 40: H.R. Lerche, Boundary Crossing of Brownian Motion. v, 142 pages, 1986.

Vol. 41: F. Baccelli, P. Brémaud, Palm Probabilities and Stationary Queues. vii, 106 pages, 1987.

Vol. 42: S. Kullback, J.C. Keegel, J.H. Kullback, Topics in Statistical Information Theory. ix, 158 pages, 1987.

Vol. 43: B.C. Arnold, Majorization and the Lorenz Order: A Brief Introduction. vi, 122 pages, 1987.

Vol. 44: D.L. McLeish, Christopher G. Small, The Theory and Applications of Statistical Inference Functions. vi, 124 pages, 1987.

Vol. 45: J.K. Ghosh (Editor), Statistical Information and Likelihood. 384 pages, 1988.

Vol. 46: H.-G. Müller, Nonparametric Regression Analysis of Longitudinal Data. vi, 199 pages, 1988.

Vol. 47: A.J. Getson, F.C. Hsuan, {2}-Inverses and Their Statistical Application. viii, 110 pages, 1988.

Vol. 48: G.L. Bretthorst, Bayesian Spectrum Analysis and Parameter Estimation. xii, 209 pages, 1988.

Vol. 49: S.L. Lauritzen, Extremal Families and Systems of Sufficient Statistics. xv, 268 pages, 1988.

Vol. 50: O.E. Barndorff-Nielsen, Parametric Statistical Models and Likelihood. vii, 276 pages, 1988.

Vol. 51: J. Hüsler, R.-D. Reiss (Editors), Extreme Value Theory. Proceedings, 1987. x, 279 pages, 1989.

Vol. 52: P.K. Goel, T. Ramalingam, The Matching Methodology: Some Statistical Properties. viii, 152 pages, 1989.

Vol. 53: B.C. Arnold, N. Balakrishnan, Relations, Bounds and Approximations for Order Statistics. ix, 173 pages, 1989.

Vol. 54: K.R. Shah, B.K. Sinha, Theory of Optimal Designs. viii, 171 pages, 1989.

Vol. 55: L. McDonald, B. Manly, J. Lockwood, J. Logan (Editors), Estimation and Analysis of Insect Populations. Proceedings, 1988. xiv, 492 pages, 1989.

Vol. 56: J.K. Lindsey, The Analysis of Categorical Data Using GLIM. v, 168 pages, 1989.

Vol. 57: A. Decarli, B.J. Francis, R. Gilchrist, G.U.H. Seeber (Editors), Statistical Modelling. Proceedings, 1989. ix, 343 pages, 1989.

Vol. 58: O.E. Barndorff-Nielsen, P. Blæsild, P.S. Eriksen, Decomposition and Invariance of Measures, and Statistical Transformation Models. v, 147 pages, 1989.

Vol. 59: S. Gupta, R. Mukerjee, A Calculus for Factorial Arrangements. vi, 126 pages, 1989.

Vol. 60: L. Györfi, W. Härdle, P. Sarda, Ph. Vieu, Nonparametric Curve Estimation from Time Series. viii, 153 pages, 1989.

Vol. 61: J. Breckling, The Analysis of Directional Time Series: Applications to Wind Speed and Direction. viii, 238 pages, 1989.

Vol. 62: J.C. Akkerboom, Testing Problems with Linear or Angular Inequality Constraints. xii, 291 pages, 1990.

Vol. 63: J. Pfanzagl, Estimation in Semiparametric Models: Some Recent Developments. iv, 112 pages, 1990.

Vol. 64: S. Gabler, Minimax Solutions in Sampling from Finite Populations. v, 132 pages, 1990.

Vol. 65: A. Janssen, D.M. Mason, Non-Standard Rank Tests. vi, 252 pages, 1990.

Vol. 66: T. Wright, Exact Confidence Bounds when Sampling from Small Finite Universes. xvi, 431 pages, 1991.

Vol. 67: M.A. Tanner, Tools for Statistical Inference: Observed Data and Data Augmentation Methods. vi, 110 pages, 1991.

Vol. 68: M. Taniguchi, Higher Order Asymptotic Theory for Time Series Analysis. viii, 160 pages, 1991.

Vol. 69: N.J.D. Nagelkerke, Maximum Likelihood Estimation of Functional Relationships. v, 110 pages, 1992.

Vol. 70: K. Iida, Studies on the Optimal Search Plan. viii, 130 pages, 1992.

Vol. 71: E.M.R.A. Engel, A Road to Randomness in Physical Systems. ix, 155 pages, 1992.

Vol. 72: J.K. Lindsey, The Analysis of Stochastic Processes using GLIM. vi, 294 pages, 1992.

Vol. 73: B.C. Arnold, E. Castillo, J.-M. Sarabia, Conditionally Specified Distributions. xiii, 151 pages, 1992.

Vol. 74: P. Barone, A. Frigessi, M. Piccioni (Editors), Stochastic Models, Statistical Methods, and Algorithms in Image Analysis. vi, 258 pages, 1992.

Vol. 75: P.K. Goel, N.S. Iyengar (Editors), Bayesian Analysis in Statistics and Econometrics. xi, 410 pages, 1992.

Vol. 76: L. Bondesson, Generalized Gamma Convolutions and Related Classes of Distributions and Densities. viii, 173 pages, 1992.

Vol. 77: E. Mammen, When Does Bootstrap Work? Asymptotic Results and Simulations. vi, 196 pages, 1992.

Vol. 78: L. Fahrmeir, B. Francis, R. Gilchrist, G. Tutz (Editors), Advances in GLIM and Statistical Modelling: Proceedings of the GLIM92 Conference and the 7th International Workshop on Statistical Modelling, Munich, 13-17 July 1992. ix, 225 pages, 1992.

Vol. 79: N. Schmitz, Optimal Sequentially Planned Decision Procedures. xii, 209 pages, 1992.

Vol. 80: M. Fligner, J. Verducci (Editors), Probability Models and Statistical Analyses for Ranking Data. xxii, 306 pages, 1992.

Vol. 81: P. Spirtes, C. Glymour, R. Scheines, Causation, Prediction, and Search. xxiii, 526 pages, 1993.

Vol. 82: A. Korostelev and A. Tsybakov, Minimax Theory of Image Reconstruction. xii, 268 pages, 1993.

Vol. 83: C. Gatsonis, J. Hodges, R. Kass, N. Singpurwalla (Editors), Case Studies in Bayesian Statistics. xii, 437 pages, 1993.

Vol. 84: S. Yamada, Pivotal Measures in Statistical Experiments and Sufficiency. vii, 129 pages, 1994.

Vol. 85: P. Doukhan, Mixing: Properties and Examples. xi, 142 pages, 1994.

Vol. 86: W. Vach, Logistic Regression with Missing Values in the Covariates. xi, 139 pages, 1994.

Vol. 87: J. Møller, Lectures on Random Voronoi Tessellations. vii, 134 pages, 1994.

Vol. 88: J.E. Kolassa, Series Approximation Methods in Statistics. viii, 150 pages, 1994.

Vol. 89: P. Cheeseman, R.W. Oldford (Editors), Selecting Models From Data: Artificial Intelligence and Statistics IV. x, 487 pages, 1994.

Vol. 90: A. Csenki, Dependability for Systems with a Partitioned State Space: Markov and Semi-Markov Theory and Computational Implementation. x, 241 pages, 1994.

Vol. 91: J.D. Malley, Statistical Applications of Jordan Algebras. viii, 101 pages, 1994.

Vol. 92: M. Eerola, Probabilistic Causality in Longitudinal Studies. vii, 133 pages, 1994.

Vol. 93: Bernard Van Cutsem (Editor), Classification and Dissimilarity Analysis. xiv, 238 pages, 1994.

Vol. 94: Jane F. Gentleman and G.A. Whitmore (Editors), Case Studies in Data Analysis. viii, 262 pages, 1994.

Vol. 95: Shelemyahu Zacks, Stochastic Visibility in Random Fields. x, 175 pages, 1994.

Vol. 96: Ibrahim Rahimov, Random Sums and Branching Stochastic Processes. viii, 195 pages, 1995.